Studies in Big Data

Volume 46

Series editor

Janusz Kacprzyk, Polish Academy of Sciences, Warsaw, Poland
e-mail: kacprzyk@ibspan.waw.pl

The series "Studies in Big Data" (SBD) publishes new developments and advances in the various areas of Big Data- quickly and with a high quality. The intent is to cover the theory, research, development, and applications of Big Data, as embedded in the fields of engineering, computer science, physics, economics and life sciences. The books of the series refer to the analysis and understanding of large, complex, and/or distributed data sets generated from recent digital sources coming from sensors or other physical instruments as well as simulations, crowd sourcing, social networks or other internet transactions, such as emails or video click streams and others. The series contains monographs, lecture notes and edited volumes in Big Data spanning the areas of computational intelligence including neural networks, evolutionary computation, soft computing, fuzzy systems, as well as artificial intelligence, data mining, modern statistics and operations research, as well as self-organizing systems. Of particular value to both the contributors and the readership are the short publication timeframe and the world-wide distribution, which enable both wide and rapid dissemination of research output.

More information about this series at http://www.springer.com/series/11970

Alan Said · Vicenç Torra

Editors

Data Science in Practice

 Springer

Editors
Alan Said
University of Skövde
Skövde, Sweden

Vicenç Torra
University of Skövde
Skövde, Sweden

ISSN 2197-6503 ISSN 2197-6511 (electronic)
Studies in Big Data
ISBN 978-3-030-07374-9 ISBN 978-3-319-97556-6 (eBook)
https://doi.org/10.1007/978-3-319-97556-6

This Springer imprint is published by the registered company Springer Nature Switzerland AG
The registered company address is: Gewerbestrasse 11, 6330 Cham, Switzerland

Foreword

In 2001, William S. Cleveland has been the first to define *data science* as a new field of study, in his Bell Labs technical report intended as an *action plan* for the practicing data analyst.[1] More than a decade later, society has embraced this call for experts who combine a strong mathematical background with a solid understanding of computer science.

Data scientists are experts in machine learning and its mathematical underpinnings *as well as* the computer science necessary to process data at scale. To illustrate the challenge to train those experts, a majority of the recent Turing Awards has recognised breakthrough ideas that are crucial in understanding data science. Leslie Valiant (2010) and Judea Pearl (2011) received this honour for their contributions to the theory of computation and learning; Leslie Lamport (2013) for distributed and concurrent systems; Mike Stonebraker (2014) for modern database systems; and, of course, we should not overlook the invention of the Web by Tim Berners-Lee (2016). This year, 2018, David Patterson has been lauded for his contributions to computer architecture. Yes, the hardware itself is important too: at Google, for example, Patterson helped design Tensor Processing Units (TPUs) that enable 15–30x faster execution of machine learning algorithms at orders of magnitude lower energy consumption [1]. We can only conclude that the data scientist needs a solid foundation to grasp the key concepts in all of these sub-areas of maths and computing.

Myself a researcher with a background in data management and information retrieval, I have long been intrigued by the idea that data powers insight to help improve science and society. I recall being excited by the wonderful bundle of essays titled 'The Fourth Paradigm: Data-Intensive Scientific Discovery',[2] edited by Microsoft Research, that showcases a kaleidoscope of scientific progress enabled by the use of computers to gain understanding from data created and stored

---

[1] https://web.archive.org/web/20060111162626/http://cm.bell-labs.com/cm/ms/departments/sia/doc/datascience.pdf.

[2] https://web.archive.org/web/20091223044640/http://research.microsoft.com/en-us/collaboration/fourthparadigm/4th_paradigm_book_complete_lr.pdf.

electronically. But, the impact of data science reaches far beyond science itself. Can you name one organisation, public or private, that is not looking to hire data scientists?

The fact that I highlighted scientific contributions to our area from three different *industry laboratories* in the introduction of this academic textbook is not a coincidence—the immediate value of hands-on experience necessary to be successful in this new domain is such that it really is the industry that pushes us forward, asking us to deliver graduates that develop smoothly into capable data scientists. Already five years have passed since the Harvard Business Review put the spotlight on this exciting field[3]—*and* predicted a shortage of qualified people! Higher education, however, has not reached a definite answer to the question what should be the curriculum of the data scientist, or even where it should be taught in the institution [2].

The book you have in front of you is a very welcome contribution to resolve this situation that needs a response so urgently. Grounded in the data science master's programme offered at the University of Skövde, the authors cover the topics that every data scientist should be intimately familiar with. I especially appreciate that the book explores both theory and practice; it does not ignore the societal and organisational context the data scientist will work in and includes ample material to develop practical skills—exactly what has been missing in curricula in the past. I believe this to be the main motivation for Cleveland to define data science as a new field, and I expect that students mastering this book have not only acquired a future-proof foundation to follow developments in this fast-pacing area of study but at the same time will be ready to apply their analytic skills in real-life problems.

Now read this book cover to cover, develop your programming skills, and find yourself ready to help shape this bright future that realises the promise of data science!

Nijmegen, The Netherlands   Arjen P. de Vries
May 2018

References

1. Jouppi, N. P., Young, C., Patil, N., Patterson, D., Agrawal, G., Bajwa, R., et al. (2017). In-datacenter performance analysis of a tensor processing unit. *SIGARCH Computer Architecture News,45*(2), 1–12.
2. Berman, F., Rutenbar, R., Hailpern, B., Christensen, H., Davidson, S., Estrin, D., et al. (2018). Realizing the potential of data science. *Communications of the ACM,61*(4), 67–72.

---

[3] http://hbr.org/2012/10/data-scientist-the-sexiest-job-of-the-21st-century/.

Contents

Chapter 1
Data Science: An Introduction

Alan Said and Vicenç Torra

Abstract This chapter gives a general introduction to data science as a concept and to the topics covered in this book. First, we present a rough definition of data science, and point out how it relates to the areas of statistics, machine learning and big data technologies. Then, we review some of the most relevant tools that can be used in data science ranging from optimization to software. We also discuss the relevance of building models from data. The chapter ends with a detailed review of the structure of the book.

1.1 Introduction

Science has as its goal to explain the universe. This explanation is intended to be objective, built from observation, and suitable to base predictions on. Reproducibility of the experiments[1] is a cornerstone in science and an essential part of the scientific method itself.

As Richard C. Brown puts it, "Scientists, by careful observation and rational reflection, accumulate evidence and formulate theories in order both to explain known and to predict new phenomena" ([2] p. 5).

Data science is the science of data. Its goal is to explain processes and objects through the available data. The explanation is expected to be objective and accurate enough to make predictions. The ultimate goal of the explanations is to make informed decisions based on the knowledge extracted from the underlying data.

---

[1] The difficulty of reproducing/replicating some experiments is a cause of concern in the scientific literature. See e.g. [1].

A. Said (✉) · V. Torra (✉)
University of Skövde, Skövde, Sweden
e-mail: alansaid@acm.org

V. Torra
e-mail: Vicenc.Torra@his.se; vtorra@ieee.org

© Springer International Publishing AG, part of Springer Nature 2019
A. Said and V. Torra (eds.), *Data Science in Practice*, Studies in Big Data 46,
https://doi.org/10.1007/978-3-319-97556-6_1

1.2   Related Areas

Data science has a strong connection with other fields and can be seen as a way to integrate these. Although often discussed, there is a consensus that these fields include: statistics, machine learning, and big data technology. Let us discuss these connections briefly.

- **Statistics**. Statistics has a similar goal of analyzing data and make inferences from data. Recall the distinction between descriptive and inferential statistics. Some (as e.g. Wu [6]) even say that data science is just a new term used for traditional statistics.
- **Machine learning and data mining**. Definitions of artificial intelligence (AI) are rooted since the origin of the field in the 1940s on the idea that systems need to learn. Machine learning is the field of AI devoted to these studies. Some machine learning methods can be used to build models from data (i.e., to build explanations) and to make predictions (classification and regression problems). Some of the tools developed within machine learning are closely linked or overlap with statistical methods.
- **Big data and database technologies**. While the size of a database is not necessarily a crucial aspect for building a model and making a decision, the relevance of data science in business lies in the fact that data is pervasive, and terabytes of data are stored for their analysis. Then, methods need to be implemented. This means that algorithms need to be efficiently programmed and lead to solutions in reasonable time. To achieve this, software developers need to master the technologies for big data.

While there is a reasonable consensus on the fields that data science integrates, there is no full agreement on the names. Data mining is often used instead of machine learning, and hacking skills or just computing is used for the latter. Considering machine learning and data mining as a part of computer science, and considering computer science as a broader field than just what is required for data science, we prefer not to use this term here.

On the top of the three fields mentioned above, as data science is about to build explanations and make predictions, any data scientist needs expert knowledge. E.g., it may be inappropriate to build models for pharmacological responses based on biomarkers without any prior knowledge in the area. The same applies if the area of application is in business data. Because of that, some add expertise or expert knowledge to the three items above.

1.3 Tools

In the previous section we have connected data science with the three related areas of statistics, machine learning, and big data technologies. In this section we will list some of the tools that are commonly used within data science.

- **Optimization**. Quite a few methods for modeling can be formulated in terms of an optimization problem [4]. That is, there is an objective function to be maximized (or minimized) and a set of constraints to be considered. The goal is to find an object or a combination of objects that satisfy the constraints, and are optimal in terms of the objective function. Optimization methods study approaches to solve this type of problems. Metaheuristics is a related area, and is about finding good heuristics to solve effectively optimization problems.
- **Probability theory**. Quite a few ways to model data are based on probability theory. Graphical models, and Bayesian networks, are some of them.
- **Linear algebra**. A simple multivariate linear regression model can be better (or more easily) represented and solved using matrices and vectors, and solved using linear algebra. Optimization problems are typically formulated using linear algebra. E.g., linear equality constraints are represented as the product of a matrix and a vector of variables equal to a vector. Some other machine learning and statistical models are also represented and solved (at least for some instances) using linear algebra. This is the case of support vector machines.
- **Graphs**. Some of the information available is conveniently represented in terms of graphs. This is the case of social networks. Graph theory provides concepts and tools to analyze this type of data. Complex networks is the term to denote a network with non-trivial topological structure. Trees are also graphs with the constraint that they should not contain cycles. In addition, some of the tools for data modeling, as the graphical models, also rely on graphs for the representation of knowledge.
- **Topology**. The field of topological data analysis [3, 5] has emerged recently as a way to extract relevant characteristics from data. Chazal and Michel [3] outlines a pipeline that stresses the role of topology and geometry in the analysis. This pipeline consists of (i) input data consisting on a finite set of points coming with a notion of distance; (ii) a "continuous shape" is built on top of the data: this results into an structure over the data; (iii) topological and geometric information is extracted from the structures; (iv) the topological and geometric information are the output of the approach and correspond to the new features of the data.
- **Visual analytics**. It is difficult to understand big data. Data visualization provides tools for a more effective understanding of the data, and visual analytics additionally provides tools for analyzing large data sets and helping in decision making processes.
- **Programming languages and software**. Appropriate programming languages for big data include R, Scala, and Python. Programming language frameworks commonly used in this field include Apache Spark, MapReduce, Hadoop, Flink,

to mention a few. Specific toolkits for data visualization tools include, e.g. Tableau and Spotfire.

- **Other mathematical tools**. The type of data we can encounter can be of any type: from images to time series, and from documents to data for weather forecasting. Because of that, the type of tools we may need to analyze data can be very diverse. Just to consider these four types mentioned, depending on the application we may need tools for image processing, for time series analysis and forecasting, for natural language processing (and possibly ontologies), and tools from fluid dynamics. Given the variety of the application domains, the tools are varied and often very tailored to a specific problem or scenario.

1.4 On Models

One of the major questions in data science projects is how to build models for explaining data and being able to make accurate predictions. For this, we use machine and statistical learning.

Models are abstractions, they are typically built to establish relationships between variables and features. There are quite a few types of models. For example, statistical models, logical models, and models based on differential equations. Statistical and machine learning provides tools for learning models (model determination) from the data itself.

Some argue that the process of defining a model and evaluating it needs to be in agreement with the scientific method. This is an iterative process, where we observe and gather data, we formulate a question, formulate an explanatory hypothesis, test the hypothesis, draw conclusions, and finally take an informed action. Iteration is due to the fact that the test and its conclusions may require the reformulation of the hypothesis.

1.5 The Structure of This Book

This book gives an overview of the area of data science from a computer science perspective. We have structured the book into three parts. The first past includes the core concepts of data science, in the second part we focus on application domains, and in the third and last part we focus on specific tools for data science.

In the part covering core concepts, the first chapter is an introduction to artificial intelligence, briefly describing its major four subfields. This includes machine learning and tools for reasoning under uncertainty. The chapter finishes with a discussion of some ethical dilemma we find in relation to artificial intelligence. Some of them, are applicable to data science in general.

The next chapter dives deeper into machine learning. It presents a concise overview of machine learning methods. It includes sections on supervised machine learning (with regression and classification methods) and on unsupervised machine learning methods. There is also a description of some methods for neural networks and deep learning. Issues related to model evaluation and dimensionality reduction are also described.

The second part on application domains includes three chapters. The first one is about information fusion. Information fusion is the process of linking and combining of information into a unified representation for the purposes of decision making. Information fusion uses machine learning methods and exploits tools from reasoning under uncertainty (as information is usually uncertain).

Then, there is a chapter on information retrieval and recommender systems. The chapter describes two of the most common end user applications of data science techniques. Information retrieval is the primary driver behind modern online search engines whereas recommender approaches are found in e.g. streaming services for music and video as well as on online shopping portals.

The next chapter covers business intelligence. Data science is applied to business data in order to make sense of what is happening in a business organization. The ultimate goal is to make informed decisions taking as much advantage as possible of the data available.

The last part of the book, on tools for data science, starts with a chapter on data privacy. It gives a brief account of common privacy models (computational definitions of privacy), and some data protection mechanisms to achieve appropriate levels of privacy.

The second chapter of this part is on visual analytics. The chapter explains the importance of information visualization and visual analytics within data science. We review perceptual and cognitive aspects, as well as design and evaluation method-ologies.

The chapter on complex data analysis follows. Complex data analysis refers to data that do not fit into entity-attribute-value model. We focus on three types of data: text, images and graphs (as a way to model e.g. social media). We present examples of tools that can be used for these type of data to extract relevant information.

The last chapter of the book focuses on applied data science in the form of an introduction to Apache Spark. This is a Big Data programming framework with many applications in data science. We describe core aspects of this framework and challenges of parallel and distributed computing. Some examples using Scala are provided.

References

1. Baker, M. (2016). 1,500 scientists lift the lid on reproducibility. *Nature, 533*(7604), 452–454 (26 May 2016)
2. Brown, R. C. (2009). *Are science and mathematics socially constructed? A mathematician encounters postmodern interpretations of science*. World Scientific
3. Chazal, F., & Michel, B. (2017). An introduction to topological data analysis: fundamental and practical aspects for data scientists. arXiv:1710.04019v1

4. Luenberger, D. G., Ye, Y. (2008). *Linear and nonlinear programming*. Springer
5. Tierny, J. (2018). Introduction to topological data analysis, UPMC, LIP6.
6. Wu, C. F. J. (1997). Statistics = Data Science? (PDF). Retrieved February 23, 2018, from https://www2.isye.gatech.edu/~jeffwu/presentations/datascience.pdf

Part I
Concepts

Chapter 2
Artificial Intelligence

Vicenç Torra, Alexander Karlsson, H. Joe Steinhauer and Stefan Berglund

Abstract This chapter gives a brief introduction to what artificial intelligence is. We begin discussing some of the alternative definitions for artificial intelligence and introduce the four major areas of the field. Then, in subsequent sections we present these areas. They are problem solving and search, knowledge representation and knowledge-based systems, machine learning, and distributed artificial intelligence. The chapter follows with a discussion on some ethical dilemma we find in relation to artificial intelligence. A summary closes this chapter.

2.1 Introduction

The term Artificial Intelligence (AI) was first used in 1955 when J. McCarthy prepared a proposal to organize in summer 1956 the first meeting of researchers working on this research area: The "Dartmouth Summer Research Conference on Artificial Intelligence". The roots of the area are older than that. At that time, there were already works on models for neural networks, search and games, and machine learning.

There is no definition of artificial intelligence accepted for all. Instead, there are different points of view. We can begin recalling H. A. Simon [20] statement "The moment of the truth is a running program". Thus, AI is about building programs. Then, the discussion can be focused on the goals of these programs.

A well-known classification of competing definitions on what artificial intelligence is was given by Russell and Norvig [17]. They consider two dimensions. The first one is about the ultimate objectives of the program: either we are interested in the results of the program (its behavior or how the system acts), or we are interested in how these results are obtained (the reasoning or way of thinking). The second dimension is related to the measurement of performance or correctness. That is, related to how we can establish that the program is achieving its goals. We can compare with people, or we can establish an ideal correctness. We use rationality to refer to this ideal objective. The two dimensions define four types of systems.

V. Torra (✉) · A. Karlsson · H. J. Steinhauer · S. Berglund
University of Skövde, Skövde, Sweden
e-mail: Vicenc.Torra@his.se; vtorra@ieee.org

© Springer International Publishing AG, part of Springer Nature 2019
A. Said and V. Torra (eds.), *Data Science in Practice*, Studies in Big Data 46,
https://doi.org/10.1007/978-3-319-97556-6_2

- **Acting humanly**. The Turing test is a paradigmatic definition of artificial intelligence under this assumption. McCarthy's proposal for the Darmouth meeting includes a similar definition.
- **Thinking humanly**. Here artificial intelligence wants to reproduce the way people think. Most work on cognitive models fall in this area. See e.g. [23].
- **Thinking rationally**. Models are built focusing on what is correct. Logics is paradigmatic of this approach. Logical inference establishes what can be inferred and what cannot. Similarly, in decision theory we have the maximum expected utility principle.
- **Acting rationally**. In this case models focus on the behavior and their performance is based on a correctness measure. The rationale of this approach is that performance should not be compared with human beings, as humans err. So, systems consider an independent and objective measure that wants to be maximized. The book on artificial intelligence by Russell and Norvig [17] follows this approach.

These different approaches to artificial intelligence compete with each other. It is clear that humans make errors, and thus acting rationally is different than acting humanly. Similarly, black-box approaches (as neural networks and deep learning) can lead to good results but they process data in a quite different manner than systems following a cognitive approach. This may cause that even in the case that a black-box approach is better with respect to performance, we prefer systems that make inferences in a more human way. For example, medical support systems that make diagnoses following a cognitive approach can explain the reasons of the diagnoses. These explanations can be used by physicians to make a more informed decision.

2.1.1 Is Artificial Intelligence Possible?

The field of artificial intelligence is based on the assumption that this question can be answered in a positive way, at least in some extent. Thus, this position is rooted on materialism.

The most important argument in this discussion is probably the Chinese room argument. John Searle described his position in [18], where he distinguished between weak AI and strong AI. Strong AI, the one that he considers impossible, is not only about behavior but also about understanding. For example, Searle states the following: "According to weak AI, the principal value of the computer in the study of the mind is that it gives us a very powerful tool. (...) But according to strong AI, the computer is not merely a tool in the study of the mind; rather, the appropriately programmed computer really is a mind, in the sense that computers given the right programs can be literally said to understand and have other cognitive states".

However, assuming that artificial intelligence is possible, there are also different points of view on how artificial intelligence can be built.

Most research in the field of artificial intelligence is based on the physical symbol system hypothesis. This hypothesis was formulated by A. Newell and H. A. Simon in a work published in 1976 [15]. The hypothesis reads as follows: "A physical symbol system has the necessary and sufficient means for general intelligent action".

So, this means that we can build artificial intelligence solely processing symbols. Nevertheless, as this processing can be expensive from a computational point of view (in short, too many options to consider), they added a second hypothesis. It is called the heuristic search hypothesis. It reads as follows: "The solutions to problems are represented as symbol structures. A physical symbol system exercises its intelligence in problem solving by search—that is, by generating and progressively modifying symbol structures until it produces a solution structure".

This position is discussed by people working on biologically-inspired models. For example, those working on neural networks and deep learning, where symbols are not explicit in a system. Systems can be designed so that they react and make conclusions using symbolic data as input and output, but internally there are no symbolic structures. In neural networks, knowledge is distributed within the system. We are not able to isolate concepts within the network.

Another trend in artificial intelligence that discusses the physical symbol hypothesis comes from people working on emergent intelligence [3]. They discuss the fact that a single physical symbol is necessary and sufficient. They advocate that intelligence emerges from the interactions of independent agents. A parallelism is made with the mind as a set of specialized functional units.

Another related point of view is the one that considers that intelligence needs situatedness and embodiment [3]. That is, we need systems to sense and act the world (the real world) and not pieces of software independent of it. This line of research opposes low reactive systems to higher level reasoning systems. In other words, we may have a system with a few rules that control the effectors of a robot based on stimuli from sensors, and no symbolic reasoning. The social embedding [4], is a related theory that claims that the situatedness in terms of sensors and actuators is not required. Systems need interaction with the environment and socialization, and that is enough because the conceptual world is already situated.

2.1.2 Areas of Artificial Intelligence

Artificial intelligence has been structured from the beginning of the field into four different areas. We will enumerate them here, and then we will develop each of them in the next sections. They are the following ones.

- **Problem solving and search**. In order to solve problems effectively, we formulate them in terms of an initial state (where the system is), a goal test (that given a state informs if the goal is satisfied), and a set of operators (that permit to transform a state into another one). Then, given a problem formulated in this way, search algorithms are to find a set of operators that from the initial state lead to another one that satisfies the goal conditions.

- **Knowledge representation and knowledge-based systems**. Systems need to represent information and knowledge. There is an area that focuses on this, and also on building systems that use knowledge extensively. This type of systems are knowledge-based systems.
- **Machine learning**. An important characteristic of intelligent systems is that they are able to learn. Machine learning is the area of AI that focuses on this problem: how to use previous experiences to make systems perform better. Machine learning is also used to build systems from data.
- **Distributed artificial intelligence**. Tasks to achieve intelligence are time consuming, because of that methods have been developed that exploit parallelism and distributed architectures. In addition, there are problems that can only be solved in a distributed manner, because it is not possible to build a centralized system as there is no central authority with all the data, information, and decision power. Multi-agent systems permit to formulate and solve the latter. Distributed artificial intelligence studies these types of systems.

These four core areas are tightly related with four other areas in which artificial intelligence plays an important role. In fact, some think that they are also a fundamental part of AI. These areas are the following: (i) natural language, (ii) computer vision, (iii) robotics, and (iv) speech recognition. We will not discuss them in this chapter. For data science, natural language processing is the most relevant one. Observe that a large portion of currently available data is text. See e.g. all posts in social networks. A brief account of an application of text analysis to topic modeling is given in Sect. 2.2 of [1] (a chapter in this book).

2.1.3 Innovative Applications of AI

A large number of applications using AI techniques have been developed in the last 50 years. Games have been considered a testbed for artificial intelligence and for search algorithms. In fact, the first mechanisms for playing chess predates AI. The most recognized game players are Chinook for Checkers, that won the Man-Machine World Championship in 1994, Deep Blue for Chess, that won G. Kasparov in 1997, and AlphaGo for Go, that beat L. Sedol in 2016.

In the area of knowledge-based systems, first expert systems were developed in the 1970 s. Mycin and Prospector being among the first ones. Since then, there are a large number of deployed systems that make decisions and recommendations based on AI techniques. In 1987 the first automatic train was deployed in Japan. It was operated with a fuzzy rule-based system. That is, a rule base system in which the terms that appear in the rules were described by means of fuzzy sets (see Sect. 2.3.4). Nowadays, there are automatic trains running in several cities in the world. The current major challenge in automatic driving is about self-driving cars. In 2005 the

DARPA Grand Challenge demonstrated that a car, Stanley, was able to complete a 212 km drive in the Mojave desert (USA). Now, we have the Tesla autopilot in the market. Most car companies are working to provide cars with self-driving options.

2.2 Problem Solving and Search

The area of search provides tools to solve problems in an abstract way. That is, we establish some principles that can be used to solve a large number of different problems. Then, algorithms have been developed that permit to solve any type of problem that can be represented according to these principles.

We will consider four examples for this purpose: the shortest path, the n-queen problem, a scheduling problem, and a symbolic integration problem. The shortest path is about finding the shortest path between two points in a graph (e.g., find the way with minimal distance from a city a to another b in a map). The n-queen problem consists on placing n queens in a chess board of dimension $n \times n$. Naturally, the queens should be placed so that they do not attack each other (i.e., we should not have two queens in the same row, column, or diagonal). Our scheduling problem consists of assigning professors to rooms for teaching taking into account a set of constraints. Constraints are about the number of lecturing hours, hours per course, and e.g. that professors can only give one lecture at a time. The integration problem is about finding the solution of the integral of a function. The integral is computed symbolically and not numerically (i.e., we look for a mathematical expression that is the integral of another given mathematical expression).

In order to solve a problem we consider states (possible configurations of the world) and then solving the problem is to move from the initial state to a state that satisfies our objectives. An important aspect is therefore to define what a state is. In the case of the shortest path, we can define the state as being in a certain town. In the case of the n-queen problem we can define a state as a set of queens (less than or equal to n) and their position with the requirement that they do not attack each other. In the scheduling problem, a state can be a list with the lectures assigned to a subset of all professors.

Then, we can represent a problem to be solved in terms of the following elements:

- **An initial state**. It is usual to consider as the initial state the state where the system finds itself before running the algorithm. In the case of building the shortest path between towns a and b, it corresponds to be in town a. In the n-queen problem, we can define it as no queen on the board. Similarly, in the case of the scheduling problem, we can define it as all professors have no teaching duties. In the case of the integral problem, we have the expression to be integrated and we have not performed any operation on that expression.
- **Operators**. They are the options that the system can consider at a given state. They permit to transform a state into another state. In the case of the shortest path, as a state is a given town, we usually consider as operators the fact of moving to the

next town (that is directly connected in a map). In the case of the n-queen problem, an operator is adding one additional queen in the board in such a way that does not attack the ones already there. This is usually done at a row-level. Assuming that we have already n queens in the first n rows, we will consider adding the new queen in the row $n + 1$ at any of the n columns (checking that the position does not attack queens already on the board). Similarly, in the case of the scheduling problem, operators assign a professor a lecture in a room. In the case of integrating an expression, we would have as operators those procedures we study in calculus that permit to integrate some basic expressions (e.g., integration of polynomials, of trigonometric functions, etc.).

- **Goal test**. We need a function that tests whether a state is a goal state or not. In some cases, as in the case of the shortest path there is only a single state that satisfies our goals (the destination). In other cases there are several states that satisfy the goals (the case of the n-queen problem). In some cases we do not even know which is the goal state and we can only test if a state is a goal state or not. This is the case in symbolic integration. The solution is an expression without the integral symbol, but we do not know which expression we are going to obtain (otherwise we would already know the solution!).

Note that the definition of states and operators are closely linked. Operators need to permit to transform one state into another, and should be able to permit to transform the original state into one that satisfies the requirements. For example, in the case of the n-queen problem we can consider a state as a board with all the queens already located (but maybe attacking each other). In this case operators would rearrange already located queens.

Once initial state, operators, and goal test are defined we can use search algorithms to solve the problem. There are a few of them. Depth-search, A* are some of them. They all proceed in a similar way.

At any point of execution, assuming that we are still unaware of the solution, there is a list of states to be considered. Initially this list is just the initial state. Then, we select one state from the list and we check if this state satisfies the conditions of a goal state. If this is the case, the search is finished. We have reached the goal. Otherwise, we apply to this state all applicable operators. In this way we get a list of states (i.e., the application of an operator to a state gives a new state), and we add this list to the ones still open for posterior analysis. Algorithms differ on how we select a state for its expansion. See e.g. [17] for details on the algorithms.

2.3 Knowledge Representation and Knowledge-Based Systems

When we develop an intelligent system, it is usual that we need to represent some background knowledge (some knowledge that is necessary for its operation and that is needed before its initial deployment), or knowledge acquired during its operation.

Knowledge needs to be accessed and to be used for inference and reasoning. As different systems have different requirements on access and inference, as well as on the type of knowledge they need to store, different knowledge representation formalisms have been defined. Most of them include a knowledge base (where we represent the knowledge at a given time) and an inference engine (that permits to compute new pieces of knowledge from the ones in the knowledge base taking into account input data as well as system's internal states).

The main types of knowledge representation formalisms are (i) logics, and languages based on logics; (ii) rules; (iii) semantic networks and frames; and (iv) neural networks and related mechanisms. An ontology is the term used for the knowledge represented in the system. The discussion of these formalisms is outside the scope of this chapter. See e.g. [12] for details on ontologies and knowledge representation formalisms.

Systems need to operate in the real world, and the information on the real world is usually uncertain and incomplete. Because of that, systems need representation formalisms that cope with this type of information. The area of reasoning under uncertainty and approximate reasoning studies and developes tools for this purpose. We review some of the main concepts in the next section.

2.3.1 Reasoning Under Uncertainty

Uncertainty can appear under different forms. Because of that, there are concepts to express different types of uncertainty. We review them in this section. We also discuss some of the tools to represent uncertainty.

We have incompleteness when our knowledge does not cover all the needs the system has on the domain. Uncertainty can appear in different flavors. It is usual to classify uncertainty in different types. The main types are randomness, ignorance, imprecision, and vagueness.

We have randomness when the outcome of an experiment is different when we repeat it. Tossing dice and coins are examples of randomness. Randomness is usually modeled using probability measures and reasoning on tools based on the Bayes theorem. e.g., in the case of a fair dice, we assign a probability of 1/6 to each of the possible outcomes (i.e., 1, 2, 3, 4, 5, and 6).

We have ignorance when we have a lack of knowledge. In some applications, it is usual to model ignorance on a set of events by means of a uniform distribution on them. For example, when we take a new die we assign to each outcome a probability of 1/6, even if we do not know whether the die is fair or not. Nevertheless, it is important to distinguish ignorance and fairness. It is quite different to know that the die is fair that not knowing anything about it. Belief functions are an alternative formalism to represent ignorance, and has the advantage that permits us to represent in a unified formalism both randomness and ignorance. In the case of a die this means that we can distinguish the case that the die is fair and the case that we have no idea about the fairness of the die. This difference can not be expressed in probability distributions.

Imprecision is when a statement is made true by more than one value. It is precise, of course, when there is only one that makes true. Let us consider the statement that the temperature is larger than 15. This statement is imprecise. We have uncertainty as any value larger than 15 can be the current temperature.

Vagueness is when truth is a matter of degree. It is usual to consider that statements can only be true or false. This is also the assumption behind probability theory and randomness. In that case, we may not know the outcome of an experiment (e.g. heads or tails), but once known, any statement (as e.g. heads) is either true or false. Vague information and statements as e.g. "the station is near" or "the temperature is high" challenge this approach. The (degree of) possibility of a given distance (say 100 meters) when we know that the station is near is not true or false (formally, 1 or 0), but a value in [0,1]. Fuzzy sets are used to represent vagueness.

In the following sections we present three of the theories to represent uncertainty.

2.3.2 Bayesian Theory

Probability theory has been used extensively to model uncertainty. There exist several competing interpretations of probability (see e.g., [9] and Chap. 1 in [26]). Under the Bayesian interpretation, probabilities represent and agent's *degree of belief* regarding an *uncertain* outcome X, i.e., a variable of interest. Let us assume that the state space for X is denoted by Ω_X and also assume that specific instantiations of X are denoted by x (i.e., $x \in \Omega_X$).

One way to measure degrees of belief is considering the prices one is willing to pay for gambles (see e.g., [5]). That is, one can think of an agent's degree of belief regarding the instantiation $x \in \Omega_X$ as the fair price $p(x)$ at which one is both willing to buy and sell. Let $g_X(X)$ be a function we call *gamble* defined in terms of the true outcome in the following way:

$$g_x(X) \triangleq \begin{cases} 1, & \text{if } x \text{ is the outcome of } X \\ 0, & \text{otherwise} \end{cases}.$$

Given the information represented in this function, it would be unreasonable to buy $g_x(X)$ at a more expensive price than 1, since that would incur a *sure loss* (due to the fact that the maximum return of the "investment" cannot be higher than 1) and it would also be unreasonable to sell $g_x(X)$ for less than 0 since one would in that case end up with a debt (in comparison to at least a reward of 0 from the output of the gamble). This line of reasoning provides us with the traditional boundaries for probabilities, i.e., $p(x) \in [0, 1]$, $x \in \Omega_X$. Furthermore by the assumption that the state space Ω_X is *mutually exclusive*, i.e., only one element of Ω_X can be the true instantiation of X, it is unreasonable to have an aggregated buying price of gambles

$g_{x_1}(X) + \cdots + g_{x_n}(X) > 1$, $|\Omega_X| = n$, since one would then effectively have a net negative reward. This line of reasoning summarises the following axioms of degree of belief as represented by a *probability distribution*:

$$p(x) \geq 0 \quad \text{for all} x \in \Omega_X$$
$$p(x) \leq 1 \quad \text{for all} x \in \Omega_X$$
$$\sum_{x \in \Omega_X} p(x) = 1.$$

The most important feature of Bayesian theory is its mechanism to update an agents' belief when a new *observation* regarding the variable of interest becomes available, denoted as *Bayesian updating*.

Bayesian Updating

In order to infer an updating mechanism within Bayesian theory, we will use that for two variables X and Y, with state spaces Ω_X and Ω_Y, we have the following relations:

$$p(X|Y) \triangleq \frac{p(X, Y)}{p(Y)} \tag{2.1}$$

$$p(Y|X) \triangleq \frac{p(X, Y)}{p(X)}, \tag{2.2}$$

where $p(X|Y)$ denotes the *conditional probability* of observing instantiations of X given that we have certain instatiations of Y (and the other way around for $p(Y|X)$) and where $p(Y, X)$ denotes the *joint probability* of observing certain joint instantiations of X and Y. Hence, given that we have made the observation $y \in \Omega_Y$ as the true value of Y, i.e. we instantiate Y with y, we can infer the following from Eqs. (2.1) and (2.2), also known as *Bayes' theorem*:

$$p(X|y) \triangleq \frac{p(y|X)p(X)}{p(y)},$$
$$= \frac{p(y|X)p(X)}{\sum_{x \in \Omega_X} p(y|x)p(x)},$$

where $p(X)$ is referred to as the *prior distribution*, i.e., the distribution that reflects an agents' degree of belief before the observation y was made, and where $p(y|X)$ is the *likelihood function* reflecting the likelihood of observing y for certain values of X. Note that the likelihood is not a probability distribution since:

$$\sum_{x \in \Omega_X} p(y|x) \neq 1 .$$

The distribution $p(X|y)$ is denoted as the *posterior distribution* and is the result of updating the prior given the observation y into the new distributions that reflects the agent's new updated degree of belief.

As an example, assume the state space $\Omega_X = \{x_1, x_2\}$ for a random variable X of interest and that some information are available regarding the true state in the form of another random variable Y, e.g., a measurement, with a state space $\Omega_Y = \{y_1, y_2, y_3\}$. Given a uniform prior:

$$p(x_1) = 0.5,$$
$$p(x_2) = 0.5,$$

and a likelihood function:

$$p(y_1|x_1) = 0.8, \ p(y_2|x_1) = 0.1, \ p(y_3|x_1) = 0.1$$
$$p(y_1|x_2) = 0.1, \ p(y_2|x_2) = 0.8, \ p(y_3|x_2) = 0.1,$$

we observe that y_1 constitute an increase in degree of belief for x_1, y_2 for x_2 while y_3 does not change the posterior since using the Bayes' rule we obtain the following results:

$$p(x_1|y_1) = 0.89, \qquad\qquad p(x_2|y_1) = 0.11$$
$$p(x_1|y_2) = 0.11, \qquad\qquad p(x_2|y_2) = 0.89$$
$$p(x_1|y_3) = 0.5, \qquad\qquad p(x_2|y_3) = 0.5$$

2.3.3 Evidence Theory

Evidence Theory [6, 19] which is also known as Dempster-Shafer theory, like Bayesian theory, provides an ability to express uncertainty but in a different way.

The uncertainty regarding an unknown random variable X for a given frame of discernment Ω_X is presented in the form of a so called mass function m, that is defined as:

$$m : 2^{\Omega_X} \mapsto [0, 1]$$
$$m(\emptyset) = 0$$
$$\sum_{A \subseteq \Omega_X} m(A) = 1$$

where any subset A of elements of Ω_X is called a focal element whenever its corresponding mass $m(A)$ is non-zero.

In addition to the mass function m, two important functions of m have been defined that are used to measure uncertainty. They are belief and plausibility and are defined [19] as:

$$Bel(A) = \sum_{B \subseteq A} m(B)$$

$$Pl(A) = \sum_{B \cap A \neq \emptyset} m(B).$$

The belief function Bel states to what extent the evidence supports A and the plausibility Pl states to what degree the evidence does not contradict A, where $A \subseteq \Omega_X$. In the same way that there are several interpretations for what a probability is, there are different interpretations for belief functions. For one of these interpretations, belief and plausibility define an upper and a lower bound for the probability of A respectively. That is,

$$Bel(A) \leq p(A) \leq Pl(A)$$

Note that when $m(A) = 0$ for any set A with cardinality larger than 1, then $P(\{x\}) = m(\{x\})$ for all $x \in \Omega_X$ define a probability distribution and $Bel(A) = p(A) = Pl(A)$ for all A.

Let us reconsider the case of a die. Using evidence theory we can distinguish the case of a fair die and the case of absolute ignorance about the fairness of the die. We would represent a fair die establishing a mass of 1/6 to each of the possible outcomes of the die. Using $\Omega_X = \{1, 2, 3, 4, 5, 6\}$ as the frame of discernment (i.e., the possible outcomes), we would represent fairness with the following mass function m_f:

$$m_f(\{1\}) = m_f(\{2\}) = m_f(\{3\}) = m_f(\{4\}) = m_f(\{5\}) = m_f(\{6\}) = 1/6,$$

and $m_f(A) = 0$ for all other A subsets of Ω_X. In particular, we define $m_f(\{1, 2, 3, 4, 5, 6\}) = 0$. This probability defines a probability because, as discussed above, it is zero for sets of cardinality larger than 1. It is easy to check that for this mass function, we have that $Bel(A) = Pl(A)$ for all A. Note also that these functions are additive ($Bel(A \cup B) = Bel(A) + Bel(B)$ for all $A \cap B = \emptyset$) and that $Bel(\Omega_X) = 1$.

In contrast, we can represent ignorance using $m_f(\Omega_X) = 1$ and $m_f(A) = 0$ for all $A \neq \Omega_X$. Observe that in this case we can deduce that $Bel(A) = 0$ for all $A \neq \Omega_X$ but that $Pl(A) = 1$ for all $A \neq \emptyset$. Note also that neither Bel or Pl satisfy the additivity condition.

The topic of evidence theory is further developed in another chapter of this book. See [22].

2.3.4 Fuzzy Sets

In standard sets membership of an element to a set is Boolean. This means that either the element belongs to the set or it does not belong to it. Fuzzy sets relax this property. We have partial membership. In order to distinguish standard sets and fuzzy sets, we use the term crisp sets for the former.

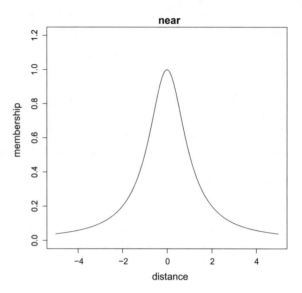

Fig. 2.1 Fuzzy set representing nearness, the reference set is $\mathbb{R}$ and corresponds to a signed distance to a reference point. It is signed as we consider whether we reach the point from the left or the right

There exist several ways to define crisp sets, one of them is in terms of characteristic functions. Given a reference set X, the characteristic function of a set $A \subseteq X$ is a function from X into $\{0, 1\}$. Naturally, $\mu_A(x) = 0$ when the element x does not belong to A and $\mu(x) = 1$ when the element x belongs to A.

Fuzzy sets are defined in terms of membership functions. They are similar to characteristic functions but in this case they are functions from X into the interval $[0, 1]$. Then we have that $\mu(x) = 0$ when the element x does not belong to the set, $\mu(x) = 1$ when the element belongs to the set, and $\mu(x)$ in $(0, 1)$ when we have a partial membership.

Fuzzy sets permit us to represent vague concepts. For example, we can define the concept near to zero in terms of the following membership function (Fig. 2.1 represents this function in the interval $[-5, 5]$). Presume that units are kilometers.

$$\mu(x) = \frac{1}{1 + x^2}.$$

This membership function shows that the maximum nearness to zero is of course when $x = 0$. In this case, $\mu(0) = 1$. Then, the farther we go either on the positive numbers or on the negative numbers, the lower the *satisfaction* of nearness. When we are at 0.5 Km from the zero, we have still a high membership (i.e., $\mu(0.5) = 0.8$) but then membership decays rapidly ($\mu(1) = 0.5$ and $\mu(2) = 0.2$).

This example illustrates that with fuzzy sets we can represent a soft transition between those elements that belong to a set and those that do not belong to the set. Observe that when we use a crisp set to represent nearness to a certain reference point, we need to establish a point from which membership is 0. Say that we establish this point to be 0.5 km, then all positions at less than 0.5 Km will be absolutely near and all positions at more than 0.5 Km will be not near. It does not matter whether we are at 0.51 km, at 1km, or at 10 Km from the point.

Fuzzy sets theory [13] studies how to operate with fuzzy sets. For example, basic questions include how to compute the union, the intersection, the complement of fuzzy sets. Then, the relation between set theory and logics has been exploited in fuzzy sets theory to develop fuzzy logic as well as the theory behind fuzzy rule based systems.

Fuzzy rule based systems are probably the most successful application of fuzzy set theory. They have been used extensively in control and modelling. A fuzzy rule based system is defined in terms of rules of the form

$$\text{If} < \texttt{antecedent} > \text{then} < \texttt{consequent} >$$

where $< \texttt{antecedent} >$ is defined in terms of conjunctions and disjunctions of expressions involving the variables of the systems and terms defined in terms of fuzzy sets.

For example, in a system to control the temperature of a device we may have a rule of the form

$$\text{If} \, \epsilon \, \texttt{is near zero and} \, \Delta\epsilon \, \texttt{is decreasing}$$
$$\texttt{then} \, \texttt{control} - \texttt{variable to zero.}$$

For details on fuzzy sets and systems, see e.g. [13]. Fuzzy systems, their construction and its application to fuzzy control is described in [7].

2.4 Machine Learning

Intelligent systems need to be built, and once built we expect them to improve their performance. Machine learning provides methods for building intelligent systems and also for improving the performance of existing systems. There is naturally a dependence on the machine learning methods and the knowledge representation formalism used by a system (e.g., learning methods for logical expressions, for rules, or for neural networks).

Methods are usually classified into three main categories: supervised, unsupervised and reinforcement learning. The categories are built taking into account the amount of information we have about what is to be learnt. Most of the methods assume that we have as input a database. We understand it as a set of examples that are used in the learning process. When the database is defined in terms of a set of records, each record is considered as an example.

In supervised learning, we have the set of examples to be used in the learning process and for each of them we have an attribute that is distinguished from the others and that we want to learn. For example, consider a bank database with data from customers loans, each loan includes a Boolean attribute informing whether the loan was paid back. Then, we can use this data to build a model for analysing new applications. In this case the distinguished attribute is the Boolean attribute, and the goal of a machine learning methods is to make e.g. a set of rules that establish when it is safe (for the bank) to accept a new application. In this example, the decision is Boolean, and thus we build a system with a Boolean output. Other systems may have other types of variables as output. Typically, we distinguish between regression and classification problems. In regression problems the output attribute is numerical, and in classification the attribute is categorical. There are quite a few methods for supervised learning. The most well known ones are decision trees and support vector machines (SVM). Neural networks and deep learning can also be seen as belonging to supervised learning.

In unsupervised learning there is no such distinguished attribute in the database. In this case, algorithms try to find patterns in the data. Clustering algorithms are an example of supervised learning. These algorithms look for sets of examples that are similar and well separated from others, and build partitions (clusters) of records or taxonomies (dendrograms) so that data scientists can visualize the similarities and differences between examples. Methods for association rule mining also belong to unsupervised learning. They are used in market basket analysis, and their goal is to find sets of items that were bought together. Methods to find latent variables as principal components and singular-value decomposition (SVD) can also be seen as belonging to unsupervised learning. There are some methods related to neural networks that are for unsupervised learning. See e.g., self-organizing maps [14].

Reinforcement learning is based on the fact that there is no complete information on the output, but only partial knowledge. More specifically, a system receives rewards and penalties in relation to its actions.

For details on machine learning, we recommend the chapter in this book [8, 10].

2.5 Distributed Artificial Intelligence

Parallel machines have influenced artificial intelligence, and new methods and approaches have been developed. We can distinguish two categories.

- **Parallel artificial intelligence**. This category encompasses those methods that exploit computational power to solve classical AI problems. This is the case of parallel algorithms for search problems and for centralized planning. That is, there is a single description of a problem and use a distributed approach to compute the solution of the problem, or a distributed system to solve the problem. In this case, a kind of central authority organizes how to find or solve the problem. The systems Deep Blue and Watson can be seen from this perspective.
- **Distributed artificial intelligence**. We model and solve a problem by means of a set of autonomous agents. Solutions are found by means of interaction and

cooperation between agents. We do not have a central authority that have all the information of a problem, neither a central authority that finds and organizes the computation of the solution. The area of multi-agent systems [27] focuses on this type of problems. They can be used for example, for problems related to decentralized optimization under uncertainty. In a decentralized optimization problem there is no central authority that knows all constraints. Instead, different companies have different constraints, and they do not want to share them. Because of that the solution should be found by means of interaction and cooperation between the agents that represent these companies. For example, through auctions and where the agents bid according to their interests. Another type of problem considered in distributed artificial intelligence corresponds to simulation (e.g. agent-based simulation).

2.6   Ethics and Artificial Intelligence

Virtually all areas of human activity are susceptible of being affected by AI and intelligent systems are today widely used by banks, hospitals, and telecom companies. They can also be found in smartphones and personal computers, in traffic regulation systems and driving autonomous cars. Since the beginning of the Industrial Revolution, automation has had in impact on the physical aspects of human work. AI is now having a similar impact on intellectual work and sophisticated cognitive tasks (as those performed by lawyers or psychotherapists) are likely to be partially, some say even entirely, substituted by artificial systems. These intelligent systems make (or assist humans to make) decisions that may have ethically significant consequences. From the banks approval of a mortgage loan to the choices made by the computer that flies an airplane, decisions made by AI systems can have serious, both positive and negative, effects on peoples lives. Since these systems work partially or totally unmonitored by humans, they should be designed and/or trained to take into consideration ethically relevant constraints. Otherwise, there is a risk that they make (or contribute to make) ethically wrong decisions [16]. Some problems generated by the use of AI systems are related to the three central ethical areas of personal autonomy, justice and maximization of wellbeing.

Autonomy and Privacy
AI systems may seriously limit our autonomy, i.e. our capacity to decide what we want to do with our lives. The principle of autonomy implies, among other things, that people have a right to choose which private information should be accessible to others. Intelligent systems, especially those processing Big Data, handle large amounts of private information as, for example, emailing, Internet surfing, geographic position, Google searches, economic transactions, home appliances, mobile phone use, etc. In order to respect personal autonomy, people must be able to give informed consent to the commercial and/or public use of the data that concerns them. However, for consent being informed, people must clearly understand the potential risks derived

from disclosing to third parties their personal data and how it may negatively affect their private and professional lives and civil rights. Today, there are serious doubts about the efficacy of the methods being used to guarantee that consumers really understand what they are giving consent to [16]. The principle of autonomy also requires that people understand that AI systems can be used to manipulate them psychologically with political and commercial purposes [11]. There are vast potential societal benefits derived from the use of private data but these have to be balanced with the risk of weakening the ethically crucial respect for personal autonomy.

Justice and Non-discrimination

Justice, as an ethical principle, implies that all humans have the same rights and duties and deserve to be treated fairly. From Google automatically labeling pictures of black people as gorillas [21] to women not getting the same quantity of advertisements of high-income jobs as men, there has been a large array of cases that show that AI systems sometimes treat people in discriminatory ways (racist, sexist, etc.). In order to avoid the problem, it ought to be possible to understand the reasons why AI systems, that in theory ought to be immune from discriminatory biases, make such prejudiced decisions. However, depending on the technology used to develop the AI systems (e.g. systems based on complex neural networks) [2], it may be nearly technically impossible to identify the causes of the problem, while in other cases the systems may have been trained using input chosen according to unconscious human biases. Creating systems that do not treat people in a discriminatory way is hence a serious technical and ethical challenge for the developers of non-discriminatory AI systems.

Maximizing Wellbeing and Social Responsibility

The use of AI systems has unmistakable societal benefits (e.g. better medical diagnoses and treatments, more secure traffic, better research, less poverty...) but it may also generate negative effects on both individual and societal wellbeing. One of the main concerns is that many employments are likely to disappear because of the use of systems able to simulate sophisticated human cognitive capacities [16]. The potential losses will affect both lowly qualified jobs (e.g. because of fully automated factories and restaurants) and highly qualified professions (e.g. in accounting, education, financial analysis...). There is also a risk of loss of ethically valuable features of some professions. Intelligent systems may outperform humans at the cognitive level but they lack genuine emotional capacities. Many professions (health care, education, etc.) have essential emotional elements, as empathy and compassion, that may disappear when people are substituted by artificial systems. Substituting a human employee by cheaper, and cognitively more efficient, intelligent systems may therefore have unexpected negative emotional consequences that have ethical relevance.

Artificial Moral Agents

The key ethical principle of non-maleficence ("do no harm") is of crucial relevance for the design of intelligent artificial agents. The AI system of a self-driving vehicle may either protect the car but damage pedestrians or do the opposite, depending on its decision-making algorithms. Therefore, the design and/or training of the algorithms that regulate the vehicles behavior have high ethical relevance. Many of the problems discussed in this section are caused by the current amorality of existing AI systems. One of the main challenges in AI research is to build artificial moral artificial agents [24, 25]. Those who are skeptical against the philosophical or technical possibility (or the genuine will) to develop such moral intelligent systems tend to be pessimistic about the benefits of AI [2] and fear the collapse of civilization if malevolent artificial general intelligence (AGI) takes over. However, some current projects [25], may indicate that Friendly AI, or Machine Ethics, is a real possibility.

2.7   Summary

In this chapter we have given an overview of artificial intelligence and its four major subfields: problem solving and search, knowledge representation, machine learning and distributed artificial intelligence. The chapter finishes with a discussion on the ethical dimensions of AI. Machine learning is discussed in more detail in another chapter of this book [8]. Some of the concepts presented here about representation of uncertainty will be resumed in the chapter [22] of this book on information fusion.

References

1. Bae, J., Karlsson, A., Mellin, J., Ståhl, N., & Torra, V. (2018). *Complex data analysis*. In A. Said, & V. Torra (Eds.) Data science in practice
2. Bostrom, N., & Yudkowsky, E. (2011). *The ethics of artificial intelligence, in cambridge handbook of artificial intelligence*. In: W. Ramsey & K. Frankish (Eds.), Cambridge University Press
3. Brooks, R. A. (1991). *Intelligence without reason*. MIT AI Memo No. 1293
4. Collins, H. W. (1996). Embedded or embodied? a review of Hubert Dreyfus' What computers still can't do. *Artificial Intelligence, 80*, 99–117.
5. Craig, E. (1998). *Routledge encyclopedia of philosophy, Routledge*. Entry: Probability theory and epistemology
6. Dempster, A. P. (1968). A generalization of Bayesian inference. *Journal of the Royal Statistical Society, Series B, 30*(2), 205–247.
7. Driankov, D., Hellendoorn, H., & Reinfrank, M. (1993). *An introduction to fuzzy control*. USA: Springer-Verlag.
8. Duarte, D., & Ståhl, N. (2018). *Machine learning: a concise overview*. In: A. Said, & V. Torra (Eds.) Data science in practice
9. Fine, T. L. (1973). *Theories of probability: An examination of foundations*. Academic Press
10. Hastie, T., Tibshirani, R., & Friedman, J. (2001). *The elements of statistical learning: Data mining, inference, and prediction* (2nd ed.). Berlin: Springer.
11. Helbing, D., Frey, B. S., Gigerenzer, G., Hafen, E., Hagner, M., Hofstetter, Y., van den Hoven, J., Zicari, R. V., & Zwitter, A. (2017). *Will democracy survive big data and artificial intelligence?*. Scientific American
12. Jakus, G., Milutinović, V., Omerović, S., & Tomažič, S. (2013). *Concepts, ontologies, and knowledge representation*. Springer
13. Klir, G. J., & Yuan, B. (1995). *Fuzzy sets and fuzzy logic: theory and applications*. UK: Prentice Hall.

14. Kohonen, T. (1982). Self-organized formation of topologically correct feature maps. *Biological Cybernetics, 43*(1), 59–69.
15. Newell, A., & Simon, H. A. (1976). Computer science as empirical inquiry: symbols and search. *Communications ACM, 19*(3), 113–126.
16. Russell, S., Dewey, D., & Tegmark, M. (2015) Research priorities for robust and beneficial artificial intelligence. *AI magazine*
17. Russell, R., & Norvig, P. (1995). *Artificial intelligence: a modern approach.* Prentice-Hall. 1st. Edition 1995. 2nd Edition 2002. 3rd Edition 2010. http://aima.cs.berkeley.edu/
18. Searle, J. (1980). Minds, brains, and programs. *Behavioral and Brain Sciences, 3*(3), 417–457.
19. Shafer, G. (1976). *A mathematical theory of evidence.* Princeton, New Jersey: Princeton University Press.
20. Simon, H. A. (1995). Artificial intelligence: An empirical science. *Artificial Intelligence, 77,* 95–127.
21. Simonite, T. (2018). When it comes to gorillas. *Google Photos Remains Blind, Wired, 01*(11), 18.
22. Steinhauer, H. J., & Karlsson, A. (2018). *Information fusion.* In A. Said, & V. Torra (Eds.) Data science in practice
23. Vernon, D. (2014). *Artificial cognitive systems—a primer.* MIT Press
24. Wallach, W. (2010). *Robot minds and human ethics: the need for a comprehensive model of moral decision making.* Ethics and Information Technology
25. Wallach, W., Franklin, S., & Allen, C. (2010). A conceptual and computational model of moral decision making in human and artificial agents. *Topics in Cognitive Science, 2*(3), 454–85.
26. Walley, P. (1991). *Statistical reasoning with imprecise probabilities.* Thomson Press
27. Wooldridge, M. (2002). *An introduction to multiagent systems.* Wiley

Chapter 3
Machine Learning: A Concise Overview

Denio Duarte and Niclas Ståhl

Machine learning is a sub-field of computer science that aims to make computers learn. It is a simple view of this field, but since the first computer was built, we have wondered whether or not they can learn as we do. In 1959, Samuel [40] proposed some procedures to build an algorithm intending to make computers play better checkers than novice players. It was an audacious goal mainly at that time when the available hardware was very limited. However, that shows the importance of machine learning since the first computers were introduced.

Nowadays, users demand computers to perform complex tasks and solve several kinds of new problems, while data are being produced from many devices (e.g., satellites, cell phones, sensors, among others). Researchers in all fields (e.g., statisticians, computer scientists, engineers, to cite some) have started the quest for making computers learn by proposing news techniques to meet the new users demands.

Data are the input of any machine learning system. Data contain examples from a given domain, and machine learning algorithms generalize the examples in the data to build mathematical models. The models can be used to predict new outputs from new examples. The data used as input in the training model are called training data.

This chapter aims to present an overview of machine learning and to serve as a road map to guide interested readers in applying machine learning to everyday problems and giving skills to become a data scientist. It is organized as follows: next section provides an overview of general issues in the field of machine learning. It also presents some examples to help readers to get through to the whole chapter. Section 3.2 discusses supervised machine learning algorithms: algorithms applied

D. Duarte (✉)
Universidade Federal da Fronteira Sul, Chapecó, Brazil
e-mail: duarte@uffs.edu.br; denio.duarte@his.se

D. Duarte · N. Ståhl
University of Skövde, Skövde, Sweden
e-mail: niclas.stahl@his.se

© Springer International Publishing AG, part of Springer Nature 2019
A. Said and V. Torra (eds.), *Data Science in Practice*, Studies in Big Data 46,
https://doi.org/10.1007/978-3-319-97556-6_3

when the training data have labels associated with every example. We present two common supervised learning techniques: regressors and classifiers. Some supervised algorithms for both techniques are also presented. The following section introduces another class of machine learning algorithms: unsupervised learning. The training data for unsupervised algorithm have no labels, so the learner aims at partitioning them into groups. Deep leaning is a new trend in machine learning and it is based on different architectures of artificial neural networks. We dedicate the entire Sect. 3.4 to these. Section 3.5 presents an overview of model assessment approaches. The model assessment is critical to validate the model regarding the quality of the prediction outputs. Another important issue in machine learning is the number of attributes in the training dataset. This issue is called dimension of the dataset. Section 3.6 presents some techniques to reduce the dimensionality to enhance the performance of the algorithms and help to visualize the data. The following section presents some final remarks about machine learning: data preprocessing (feature selection and scaling, missing values), bias, variance, over and underfitting. Finally, Sect. 3.8 concludes this chapter.

3.1 Introduction

Learning is a very complex process, and we cannot say, currently, that a computer can learn. With this in mind, our first definition of machine learning must be carefully reviewed. There is no general agreement about what is learning, however for human beings learning can be defined as (i) functionally as changes in behavior that result from experience, or (ii) mechanistically as changes in the organism that result from experience [6].

Computers are mathematical machines; thus, we have to consider learning as a computer program. Figure 3.1 presents pictorially traditional programming (a) and machine learning (b). Notice that, while traditional programming is concerned with finding the right output based on given inputs and a program, machine learning is concerned with finding a right program (later we call it a model) given a set of inputs and outputs (possibly empty). The learned program can now propose new outputs given new inputs.

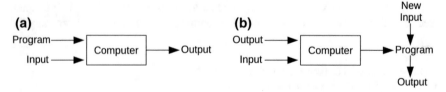

Fig. 3.1 Traditional programming (**a**) and Machine learning (**b**)

Based on Fig. 3.1, we can see how different learning means in human beings and computers. Machine learning algorithms try to create a model that represents the input to propose new outputs. This is the type of learning concerned in this chapter.

Back to the machine learning definition, we point out two definitions. The first one proposed by Samuel [40] who said that machine learning is a field of study that gives computers the ability to learn without being explicitly programmed. Remark that Samuel's definition was one of the first proposed definitions. Almost forty years later, Mitchell [32] proposed a more mathematical view of machine learning: a computer program is said to learn from experience E with respect to some task T and some performance measure P, if its performance on T, as measured by P, improves with experience E.

Example 3.1.1 Assume we want to learn whether or not a given email is spam. In this case, we give to the algorithm a set of emails S_e and divide it into two subsets: not spam (S_{nse}) and spam (S_{se}). This step is the experience (E) we give to the algorithm. Based on E, our algorithm classifies emails as spam (and consequently as not spam). This step is related to the task (T). Finally, we want to know how well our algorithm performs in classifying spam emails. This step is related to the performance of our algorithm (P). Our goal is, then, to find a T with as good P as possible. Of course, T and P depend on the quality of E, e.g., if we have a good informative S_e, our machine learning algorithm may classify all our input emails correctly.

Figure 3.2 changes a little bit of the representation of Fig. 3.1. Firstly, we call *Training Set* (X) the input of our machine learning algorithm. Depending on the task we want to accomplish, X may have a label for every example. So, X can be represented as $\{(x^{(1)}, y^{(1)}), \ldots, (x^{(m)}, y^{(m)})\}$ or $\{x^{(1)}, \ldots, x^{(m)}\}$, where m is the size of X (training set), and $y^{(j)}$ is the label of $x^{(j)}$ ($1 \leq j \leq m$). The learning algorithm (we have thousand of options) takes X as input and builds a *Model* (also known as a *hypothesis*). After assessing our model (i.e., verify how good our P is), we can feed it with new examples (X') to have predicted outputs. Take into account that, depending on our target task, the predicted outputs can be a class (discrete value), continuous values, clusters, among others. This will be further elaborated in this chapter.

Based on what we have already seen, the machine learning process can be divided into four steps: (i) get the training set X, (ii) choose and implement a learning task based on X, (iii) build a model, and (iv) assess the model with new inputs. Remark that these four steps may be repeated until we have reached a good P. Keep in

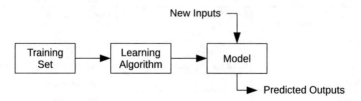

Fig. 3.2 Pictorial representation of the machine learning process

mind that when we have a machine learning problem to solve, the first challenge we face is which machine learning algorithm to use. There are thousands available, and each year other hundreds are proposed [8]. The set of possible available learning algorithms for a given machine learning problem is called *hypothesis space*. To not get lost in this huge set of choices, the learning components can be divided into three [8]:

1. Representation: a task must be represented by some algorithm. We must know what kind of learning we are interested in, and so, the hypothesis space is decreased based on the type of task.
2. Evaluation: the predicted outputs must be evaluated (assessed) to know how good the chosen representation is. Depending on the task, different evaluation functions can be used.
3. Optimization: based on the results of the evaluation component, optimization must be done. The aim of the learning algorithm is to maximize a given performance measure.

Table 3.1 presents some examples for each of the three components. For example, a *Decision Tree* may classify a training set into predefined classes. An evaluation component can be *accuracy*, i.e. how many classes are correctly classified. An *entropy* function can be used to measure the purity of the attributes within the tree, that is, the amount of information that would be needed to classify an example. Table 3.1 gives a little idea of the options we have when we design a machine learning system. Remark that we cannot pick an example for each column to design our system. Each representation has its own set of optimization and evaluation approaches. For example, we can use *Greedy Recursive Partitioning* for *Decision Trees*, and the model can be evaluated by *Accuracy*.

Although there does not exist a simple recipe to choose the best approach from one of the three components, the success (or not) of a learner depends on how well the problem is defined, as well as the quality of the training set. The former helps to find a representation that fits better to the problem, and the latter can be considered an essential *ingredient* of a learning system.

We illustrate what we have seen so far through a running example.

Table 3.1 Instances of the three components of learning algorithms

Representation	Evaluation	Optimization
Logistic regression	Accuracy	Gradient descent
Neural networks	Precision/Recall	Greedy search
Linear regression	Squared error	Beam search
Decision tree	Cost function	Linear programming
K-means	Root mean square error	Greedy steepest descent
Random forest		Stochastic approximation
Naive bayes		Greedy recursive partitioning

Table 3.2 Wind and temperature affecting pace

Wind speed (km/h)	Temperature (°C)	Pace (min)
10.5	12.3	3.5
8.9	15.4	3.2
20.2	13.7	5.5
5.10	3.1	4.0

Example 3.1.2 Table 3.2 presents an extract of a dataset about running performance (column *Pace*—the number of minutes it takes to cover a kilometer) based on the wind speed and the temperature. Suppose we want to predict a pace based on new information about the weather. X can be seen as (<10.5, 12.3>, <8.9, 15.4>, <20.2, 13.7>, <5.10, 3.1>) and y as (3.5, 3.2, 5.5, 4.0). The size of X is 4 and number of features is 2 (i.e., wind speed and temperature), and we want to predict a continuous value: the pace.

The number of features is usually known as the dimensionality of the dataset. The notion of dimensionality leads to a well-known problem in machine learning: the curse of dimensionality [21]. Considering a dataset as a set of points in a plane, the curse of dimensionality can be stated as follows: (i) learning algorithms generally work with interpolation to build models, (ii) interpolation is only possible if points are close to each other, (iii) if points are spread throughout a high dimensional space, the distance between them is large, and (iv) interpolation-based algorithms cannot build a model.

Besides, high dimensional dataset also leads to two problems: increase of computation cost and non-informative features. Suppose that we have, in our example, the following features: the running shoes price and quality. Although, quality and price are related to each other, in our dataset we can easily discard the feature price without losing essential information for our machine learning system. Notice that if we have a dataset with 200 features, discarding or merging some of them would not be an easy task. We deal with the dimensionality problem later in Sect. 3.6.

Example 3.1.3 Given Example 3.1.2 and $x' = $ <11.2, 10.0>, we want to predict a new $\hat{y}$ such that $\hat{y}$ represents a valid pace value for the wind speed and temperature given. To accomplish the prediction, we have to build a model that describes well enough our training set (X). We show how to build a model using a simple linear regression (notice that from Table 3.1 we are choosing a representation for our problem). A model can be

$$\theta_0 + \theta_1 \times x_1^i + \theta_2 \times x_2^i = \hat{y}^i$$

where θ_j are the weights or parameters (sometimes denoted by W), x_k^i represents the kth feature of the ith example in X and $\hat{y}^i$ is the predicted output of the ith example, and we want a $\hat{y}^i \approx y^i$. θ_0 represents the bias of the model (aka intercept). The challenge is to find θ_0, θ_1, and θ_2 such that our $\hat{y}^i$ value is as close as possible to y^i. Given the matrix $X_{m \times d}$ and the column vector $\Theta_{d+1 \times 1}$, we can implement our model using matrix multiplication. However, $X_{4 \times 2}$ and $\Theta_{3 \times 1}$ are not dimension-compatible. Remark that, in our representation, Θ will always have one more column in relation to X, and we can solve this problem adding a 1's column in X. As 1 is the identity element under multiplication, so, if we multiply θ_0 by 1, its value remains the same. Now, we can represent our model as:

$$\begin{bmatrix} 1 & 10.5 & 12.3 \\ 1 & 8.9 & 15.4 \\ 1 & 20.2 & 13.7 \\ 1 & 5.10 & 3.1 \end{bmatrix} \times \begin{bmatrix} \theta_0 \\ \theta_1 \\ \theta_2 \end{bmatrix} = \begin{bmatrix} \hat{y}_1 \\ \hat{y}_2 \\ \hat{y}_3 \\ \hat{y}_4 \end{bmatrix}$$

and, so, $\hat{y} = X \cdot \Theta$.

The next step is to find suitable values for the Θ. If $\Theta^T = [0, 0.5, -0.2]$ then $\hat{y} = [2.79, 1.37, 7.36, 1.93]$. Our Θ values made a fair prediction for the first example ($y_1 = 3.5$), but they failed for the others. To find fair values to Θ is an optimization problem (our third component). If we are using linear regression to represent our problem and an error function to evaluate it, we can use an optimization algorithm for finding the minimum of a function (e.g., *Gradient Descent*), and we get the following values $\Theta^T = [3.4, 0.192, -0.132]$, and, then, $\hat{y} = [3.8, 3.1, 5.5, 4.0]$. Finally, to verify if our model is performing well, we have to evaluate it (second component). For linear regression, we can use *Root Mean Squared Error* (RMSE), and the error is 0.16 (see Sect. 3.5 for details). The smallest the value of RMSE, the closer our model is to make good predictions. Given x', we have $3.4 + 0.192 \times 12.0 + -0.132 \times 11.0 = 4.2$, that is, when the wind speed is 12.0 and the temperature is 11.0, the probably pace would be 4.2.

We can evaluate our model using the same set used for training. However, the best way to test our model is against unseen examples. Later in this chapter, we describe some strategies on how to train and test machine learning models.

In this section, we presented an overview of machine learning. In the next sections, we describe approaches for implementing a range of types of algorithms to solve machine learning problems. We consider a broad classification of a learning task (i.e. machine learning algorithm): supervised and unsupervised.[1]

---

[1] There are other classes of machine learning algorithms: semi-supervised, reinforcement learning, recommender system. In this chapter, we focus on the two most popular ones. We refer [32] to the readers for classes not covered here.

3.2 Supervised Learning

Supervised learning can be applied when the dataset contains a set of labels (possibly unitary) for every example. Therefore, the dataset is divided into X, the features of the examples, and y, the labels. The labels help a learning algorithm to build the predicting model and act as a guide to the learners.

Labels may be discrete (classes) or continuous (numeric values). Depending on the type of label, we can apply classifiers or regressors. Besides, labels make the evaluation of a model easier since we have the ground-truth to compare with the predicted values. Table 3.2 shows a dataset that can be used for supervised algorithms. Remark that the label (*Pace*) represents continuous values, and, so, the dataset can be used as input for regression supervised algorithms. If we change the *pace* to discrete values (e.g., fast, slow, normal, etc.), our problem becomes a classification problem. Note that any regression problem can be turned into a classification problem by binning the continuous target values. Therefore, the first step of machine learning system design is to analyze the dataset to identify which representation must be used. In the following, we describe both regression and classification.

3.2.1 Regression

Regression is a type of supervised machine learning algorithm whose target variables are continuous values. Predicting currency exchange rates, temperatures, and the time when an event may occur are examples of regression problems since the predicted outputs are continuous values. In this section, we present some regression algorithms.

3.2.1.1 Linear Regression

When we face a machine learning problem with continuous target variables, we have to choose a representation based on regression algorithms. Linear regression is the most common algorithm used in regression and serves as the base to understand all other regression algorithms [42].

The mathematical definition of linear regression is given by the following equation:

$$h_\Theta(X) = \theta_0 + \theta_1 \times x_1^{(i)} + \cdots + \theta_m \times x_m^{(i)} \tag{3.1}$$

where θ_j are weights (θ_0 is the bias, and θ_k is the weight for the kth feature of $x^{(i)}$, $1 \leq k \leq m$), and $x_j^{(i)}$ is the jth feature of the ith example in X (dataset).

The best way to understand how linear regression works is to plot a graph with the *features* × *label*. However, most of the time the dataset is multidimensional, that is, it has more than one feature. Later in this chapter, we discuss dimensionality reduction, but, for the sake of simplicity, we consider only the feature *Temperature* from our dataset (Table 3.2). Table 3.3 shows the original dataset from Table 3.2 extended with some new examples to have more points in the graph. Figure 3.3 shows the new dataset plotted in a 2D graph ($X \times y$).

Remark that, in Fig. 3.3a, there is a line drawn by $h_\Theta(X)$ with weights that do not represent the data points very well. However, in Fig. 3.3b, the average distance of the points to the line shows that $h_\Theta(X)$ better describes the dataset. We can verify whether or not our model fits the training dataset well by measuring the (Euclidean) distances between the points and the line. In Fig. 3.3, the red lines show pictorially some distances. Remark that one way to evaluate our model is to calculate the average distance between the points and the line.

Example 3.2.1 Based on Fig. 3.3b, we have $h_{([2.3,0.222])}([10]) = 4.52$, that is, when the temperature is 10 degrees, the runner will take 4.52 minutes to run 1 kilometer. For a temperature of 12.3, our model outputs a pace of 5.03 which is higher than the ground-truth value of 3.5 (see Table 3.3).

So far, we have chosen the representation of our model (i.e., Linear Regression), and we still need to choose approaches to evaluate and optimize the model. In Sect. 3.5, we present some approaches for model evaluation. However, if we use *Mean Square Error*, our model $\Theta = [2.3, 0.222]$ gets a score of 0.641. Knowing

Table 3.3 Table 3.2 with just one feature and new examples

Temperature (°C)	Pace (min)
12.3	3.5
15.4	3.2
13.7	5.5
3.1	4.0
11.3	4.6
10.8	4.3
9.7	4.0
4.5	3.5
5.3	3.3
5.2	3.8
7.4	3.4
6.2	4.2
7.8	4.3
8.5	3.8
12.3	5.3
14.3	5.2
13.2	4.4
9.8	4.6
8.3	4.6

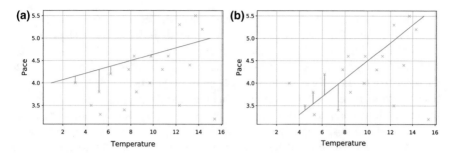

Fig. 3.3 Perform (the drawn line) of two regression models in the same dataset

that close to 0 is better, our model should be optimized. The optimization is the third component of our hypotheses space (see Table 3.1). Gradient Descent is a common solution to optimize (or train) a Linear Regression Model.

Gradient Descent updates the weights (Θ) based on a cost function. Equation (3.2) shows an implementation of a cost function.

$$J(\Theta) = \frac{1}{2m} \sum_{i=1}^{m} (h_\Theta(x^{(i)}) - y^{(i)})^2 \tag{3.2}$$

where m is the number of observations in the training dataset. Remark that $h_\Theta(x^{(i)})$ may be replaced by $\hat{y}$. The mathematical definition for the gradient descent is shown in Eq. (3.3).

$$\Theta = \Theta - \alpha \frac{\partial}{\partial \Theta} J(\Theta) \tag{3.3}$$

where α is the learning rate. Equation (3.4) shows Eq. (3.3) with the partial derivative computed for $J(\Theta)$.

$$\Theta = \Theta - \alpha \frac{1}{m} \sum_{i=1}^{m} (h_\Theta(x^{(i)}) - y^{(i)}) x^{(i)} \tag{3.4}$$

The gradient descent works as follows: (i) we randomly initialize Θ (for linear regression all Θ can be initialized to 0),[2] (ii) define a value to α,[3] and (iii) we run gradient descent until it converges. We can use $J(\Theta)$ to stop the looping. When $J(\Theta)$ stabilizes, we consider that the gradient descent has converged.

Remark that the parameters (called hyper parameters) are essential for building good machine learning systems. The ordinary linear regression needs only a good

---

[2] For some representations, zero is not a good initial value. Random values from 0 to 1 work in most of cases.

[3] Select a small value to α, say 0.01, plot $J(\Theta)$ to identify how the gradient is converging, increase α (e.g., doubling its value) up to have an expected convergence.

value for the learning rate, but it is not always the case. There are some more sophisticated algorithms that we have to choose values for all the hyper parameters needed.

Sometimes the data points cannot be described by one straight line. In this case,
we can still apply linear regression to the dataset. We just need to change our model. If
the points are not linearly organized, we can apply polynomial models (in this case, a
linear regression may be called a polynomial regression). Suppose we have a dataset
with just one feature, and the data points represent a quadratic function. We could
build a model as $h_\Theta(X) = \Theta_0 + \Theta_1 \times x_1 + \Theta_2 \times x_1^2$. It is not hard to implement
since we insert a new column representing the squared feature. Therefore, the model
becomes $h_\Theta(X) = \Theta_0 + \Theta_1 \times x_1 + \Theta_2 \times x_2$, where $x_2 = x_1^2$. Any polynomial function can be used to build a non-linear model. The characteristics of the dataset are
the guide for choosing the best one. However, the cost function must be convex to be
optimized with gradient descent (the cost function in Eq. 3.2 happens to be convex).

An alternative to training a learning model (i.e., solve the parameters Θ) is to
use normal equation. Equation (3.5) gives the mathematical definition of normal
equations.

$$\Theta = (X^T \cdot X)^{-1} \cdot X \cdot y \qquad (3.5)$$

There are two advantages of normal equation over gradient descent: there is no
learning rate and no iteration. However, the matrix that represents the dataset must
be invertible, and the computational cost to multiple and invert matrix is high.[4]

Remark that $J(\Theta)$ guides gradient descent in finding the best weights for the
model. However, sometimes $J(\Theta)$ fits well the training set ($J(\Theta) \approx 0$) but fails to
generalize the test set ($J(\Theta) \gg 0$). This situation (aka overfitting) indicates that our
model has been specialized for the training dataset, i.e., the noise in the dataset has
been taken into account during the learning process. Overfitting may happen when
the dataset is highly dimensional, and some features may be irrelevant in the training
step.

There are several techniques to combat the overfitting, the most popular is to add
regularization to the (cost) function. The regularization penalizes the function using
the weights and some other parameters (see Sect. 3.7.2 for more details).

The algorithm described here for regression problems represents a small part of
the hypotheses space for solving this kind of problem, but it can be used as the basis
for understanding other regression algorithms.

3.2.2 Classification

Classification problem is similar to a regression problem; the only difference is
the labels that are discrete values. With this in mind, a regression dataset may be
transformed into a classification dataset. We have just to discretize the values, that is,
group the continues values into classes. The label (*Pace*) in Table 3.2, for example, can

---

[4]Multiplication is $O(n^2)$, and inverse is $O(n^3)$.

Table 3.4 The dataset from Table 3.4 with Pace having discrete values

Wind speed (km/h)	Temperature (°C)	Pace	#Class
10.5	12.3	Fast	0
8.9	15.4	Fast	0
20.2	13.7	Normal	1
5.10	3.1	Normal	1

be discretized resulting in a new dataset shown in Table 3.4. The continuous values are transformed into discrete values as follows: (i) fast (*Pace* $\leq$ 3.5), and (ii) normal (*Pace* > 3.5). As we are working with mathematical representations, the classes of the labels should be represented as numerical values. Therefore, *fast* is represented by 0, and *normal* by 1 (column *#Class*). Classes may be binary (as our example) or multiclasses (we could represent the Pace as fast, regular, normal, slow, among others). In the first case, we have $y \in \{0, 1\}$, and in the second case, $y \in \{0, 1, \ldots, k - 1\}$ (where k is the number of classes). We focus this section on binary classification since they are more intuitive to understand. Besides, binary classifiers are applied to several situations: classification emails as spam or not, fraudulent transactions, problems regarding winners or losers, among others. In addition, any multi-class problem may be solved by dividing up the problem into many binary classification problems in a so-called *one* versus *all* classifications.

3.2.2.1 Logistic Regression

Logistic regression is one of the simplest and most efficient classifiers. The intuition behind logistic regression is similar to linear regression. We extend Eq. (3.1) to output $\hat{y}$ such that $\hat{y}$ in [0, 1]. Equation (3.6) presents a sigmoid (or logistic) function that always returns values between 0 and 1 inclusive and the logistic regression can be defined as $\sigma(h_\Theta(X))$.

$$\sigma(z) = \frac{1}{1 + e^{-z}} \tag{3.6}$$

Sigmoid functions behavior as follows: $\sigma(z = 0) = 0.5, 0 \leq \sigma(z < 0) < 0.5$, and $0.5 < \sigma(z > 0) \leq 1$ (Fig. 3.4 shows pictorially this behavior). Remark that we may interpret the output of Eq. (3.6) as the probability of $y = 1$ (or $y = 0$). Therefore, the binary classifier can output 1 or 0 based on a threshold defined by the user. If we want an equal distribution, the value returned by $\sigma(z)$ can be rounded, and so we have $y = 1$ when $\sigma(z) > 0.5$, or $y = 0$, otherwise.

If we use the previous cost function in the logistic regression, we would have a non-convex function which could not be optimized with steepest gradient descent. If we try it anyhow, we would risk to find a local minimum instead of the global minimum (Fig. 3.5 shows pictorially this situation: solid circles represent local minima and the open one represents the global minimum). To avoid this, we consider the cost for

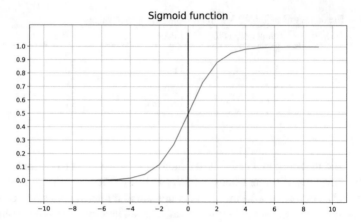

Fig. 3.4 The behavior of a sigmoid function

Fig. 3.5 Plot of a nonlinear cost function

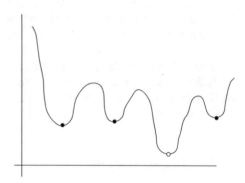

$(\hat{y}, y)$ as follows:

$$cost(\hat{y}, y) = \begin{cases} -log(\hat{y}) & \text{if } y = 1 \\ -log(1 - \hat{y}) & \text{if } y = 0 \end{cases} \tag{3.7}$$

Remark that the cost is 0 if y and $\hat{y}$ are equal to 1, and increases when $\hat{y} \to 0$. It is the behavior we want since the penalty has to increase when the distance of y and $\hat{y}$ increases. The same reasoning may be applied to when $y = 0$. Note that the Eq. (3.7) can be rewritten as $y \times (-log(\hat{y})) + (1 - y) \times (-log(1 - \hat{y}))$. Thus, we can define the cost function of linear regression as:

$$J(\Theta) = -\frac{1}{m}[\sum_{i=1}^{m} y^{(i)} log(h_\Theta(x^{(i)})) + ((1 - y^{(i)})log(1 - h_\Theta(x^{(i)})))] \tag{3.8}$$

Notice that when $y = 0$, the first part of the summation (the left side of the addition) is canceled since y multiplies the log value, and the same reasoning can be applied to the right side when $y = 1$. The gradient descent remains the same as shown in Eq. (3.3).

If we plot our dataset in a vector space, logistic regression draws a line that separates the data points based on their classes. This separation is called a decision boundary. However, sometimes the classes cannot be separated by a straight line, and in this case, one solution is to transform the feature matrix into a higher dimensional space by adding new features with higher degree. Another solution is to apply another algorithm to the problem, for example, Support Vector Machines (SVM) [15].

One-versus-all is one of the approaches to deal with multi-class classification. This approach learns one class at a time, and thus, a dataset with k classes has k sets of Θs. For example, in a dataset with three classes (0, 1, and 2), we keep the label of first class and update the others to 1; we do the same for the second class, update the others to 0, and so on. To verify in which class an example e_i belongs, we apply every learned Θ to e_i, and the highest one corresponds to the predicted class of e_i.

3.2.2.2 Decision Trees

Decision trees are another representation for solving classification problems. The model is based on decision rules implemented in the nodes of a tree. This model is more understandable for humans than logistic regression. See a pictorial representation below:

Basically, every node represents a test to be performed on a single attribute, and a child node is accessed depending on the result of the test. The testing is repeated until it reaches a leaf node, and finally the class is found. Figure 3.6 presents a decision tree for the dataset from Table 3.4. Remark that the tree covers all cases of the dataset, and we may conclude that a runner will have a normal pace under temperatures below 12.3.

The choice of the attributes for each rule and node has a major role in the success of a decision tree. The criterion is based on the information gain of the attributes (features). The attribute that gives the greatest information gain becomes the root of the tree, and the internal nodes follow the ranking of the information gain. Entropy calculates the (im)purity of an attribute regarding the classes, and it may be used to calculate the information gain (i.e. how well an attribute can describe a given class) [3]. Equation (3.9) gives the mathematical definition of an entropy for a subset X_i:

$$H(X_i) = -p_i^+ log_2(p_i^+) - p_i^- log_2(p_i^-)$$
(3.9)

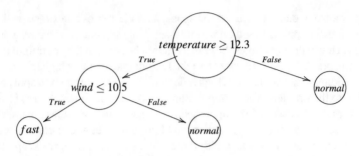

Fig. 3.6 A decision tree built from dataset in Table 3.4

where p_i^+ is the probability that a randomly taken example in X_i is positive and can be estimated by the relative frequency $p_i^+ = \frac{n_i^+}{n_i^+ + n_i^-}$; the same reasoning is used to calculate p_i^-.

The entropy for every value i of an attribute *attr* in X is calculated as follows (considering the *attr* has K different values):

$$H(X, attr) = \sum_{i=1}^{K} P(X_i) \times H(X_i)$$

where $P(X_i)$ is the probability of an example belonging to X_i and can be estimated by the relative size of subset X_i in X: $P(X_i) = \frac{|X_i|}{|X|}$. Finally, the mathematical definition of a knowing attribute *attr* information gain is given by Eq. (3.10).

$$I(X, attr) = H(X) - H(S, attr) \tag{3.10}$$

where $H(X)$ is the entropy of the whole dataset. Remark that if X is well balanced for the classes, $H(X)$ will be close to its maximum (≈ 1).

Example 3.2.2 Let X be the dataset from Table 3.4, the result from Eq. (3.9) is: $H(X) = \frac{2}{2+2} \times log_2(\frac{2}{2+2}) - \frac{2}{2+2} \times log_2(\frac{2}{2+2}) = 1$. This means that X is well balanced. It is easy to see since X is composed of two positive and two negative examples. Let t_1 be the temperature greater or equal to 12.3 and the class normal be a positive example, the entropy is: $H(X, t_1) = \frac{1}{3} \times log_2(\frac{1}{3}) - \frac{2}{3} \times log_2(\frac{2}{3}) = 0.92$. For the temperate below 12.3, the entropy is by definition 1 since there is no negative example, and logarithm of zero is not defined. The total entropy of the attribute temperature is $H(X, temperature) = \frac{1}{4} \times 1 + \frac{3}{4} \times 0.92 = 0.94$. Finally, the information gain is $I(X, temperature) = 1 - 0.94 = 0.06$. If we calculate the information gain for a wind speed less or equal to 10.5, we will have the same result. That is, both attribute have the same information gain, and so both can be the root of the decision tree. Remember that the tree is organized by following the information gain of the attributes in descending order.

The equations and formulas above consider that the attribute values are discrete, but it is not always the case: decision tree can be induced from numerical attributes as well. An approach to discretize the values follows: an attribute *attr* is sorted, the range of each class for *attr* is calculated, and the ranges are ranked by information gain. Each range corresponds to a discrete value of *attr*. A range for an attribute *attr* can be [10, 20], i.e., $attr > 10 \wedge attr < 20$.

Decision trees are the basis for lots of other tree classifiers. One of the most effective one is the random forest. Roughly, random forest trees combine into an ensemble. N random samples are selected from the dataset X; each sample is used to build a decision tree with some samples of X. Therefore, random forest trees use N decision trees to build the best model for a given dataset.

Decision trees are proposed to deal with classification problems. However, there are several approaches to adapt decision tree algorithms to regression problems [26]. A regression tree is similar to a classification tree, except that the label y takes continuous numerical values and a regression model is fitted to each node to give the predicted values of y.

3.3 Unsupervised Learning

When the dataset has no labels, that is, there is no previous classification of the examples, we apply an unsupervised learning algorithm. The goal is to infer classes or groups from the dataset without the help of the labels. In this case, the dataset is in the form $X = \{x^{(1)}, \ldots, x^{(m)}\}$. Unsupervised learning is less objective than supervised learning, since there are no labels to guide the user for the analysis. The domain of the dataset must be known by the user to build useful models, otherwise, the results may not be understandable.

Although, unsupervised learning is harder to model than a supervised learning, the importance of such techniques is growing since there are more unlabeled data than the labeled ones. Besides, many learning problems are related to unsupervised problems: recommendation, classification of customer behaviors in a website, market segmentation, among others. Clustering is the most popular technique for unsupervised learning.

Clustering is about discovering semantically related groups in an unlabeled dataset. The number of groups (aka clusters) is defined by the user based on his/her knowledge of a dataset X. For example, let's say X represents examples of heights and weights of people, and we want to separate them into 3 *T-shirts* sizes (e.g., S, M, and L). The dataset can be split into 3 clusters, and, based on the user knowledge, each cluster represents the *height* $\times$ *weight* characteristic for each *T-shirts* size.

Data clustering has been used for [18] (i) gaining insight into data, generate hypotheses, detect anomalies, and identify salient features, (ii) identifying the degree of similarity among forms or organisms, and (iii) for organizing the data and summarizing it through cluster prototypes.

K-means is one of the most popular and easy to understand clustering algorithms [19]. The basic idea is to define k centroids that help to build the clusters. Every example in the dataset will be associated to one of the k centroids. The dataset is seen as a set of data points in a plane, and the algorithm tries to group them into clusters by measuring the distance between a given data point and a centroid.

Data: $X = \{x^{(1)}, \ldots, x^{(m)}\}$, K centroids $\mu_1, \ldots, \mu_K$
Result: A set of K centroids for X
Initialize the centroids (either $K \subset X$ or data points picking from the plane);
repeat
  for *each $x^{(i)}$ in X* **do**
    $c^{(i)} = argmin_k(\sum_{k=1}^{K} ||x^{(i)} - \mu_k||^2)$; // assign $x^{(i)}$ to the closest centroids
  end
  for *each μ_i in K* **do**
    // update the centroids with the average of the points associated to them
    $\mu_i = \frac{1}{|c^{(i)}|} \sum_{x^{(j)} \in c^{(i)}} x^{(j)}$;
  end
until *K converge*;

Algorithm 1: K-means pseudo code

Algorithm 1 presents a K-means pseudo-code. The two internal loops are the main parts of the algorithm. The first one associates each data point (an example from X) to a given centroids. The second loop updates the centroids positions by averaging the points associated to them. The main loop is repeated until the data points get stable in relation to their centroids.

Figure 3.7 shows pictorially the above steps of K-means (rounded points are centroids, and diamonds represent the data points). First, three centroids are picked from the hyperplane ((a), points black, red, and yellow). Next, all data points are associated to a centroid (b), new centroids are calculated ((c), the dotted arrow shows the source and target of each centroid), and finally, the data points are updated (d). Those steps are repeated until the centroids do not change anymore. The initial centroids play an essential role in K-means algorithms, and the resulting clusters may be different depending on the initial centroids.

Although, there are methods for selecting the so-called correct number of clusters (e.g., Silhouette and CH index methods [11]), the user knowledge plays an essential role to define the number of centroids (the number of clusters). Each cluster will have a semantic meaning in relationship to the domain of the dataset for the expert in the domain.

The K-means algorithm is a clustering algorithm based on partition, i.e., the idea behind is to consider the center of data the points as the center of the corresponding cluster. Another category of clustering algorithms are those based on a hierarchy. This kind of algorithm builds a hierarchical relationship among data to cluster them. Each data point, in the beginning, is a cluster itself. The closest clusters are merged.

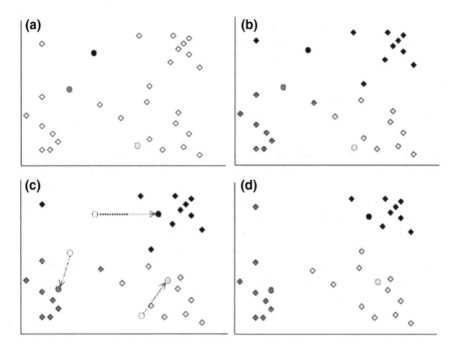

Fig. 3.7   First steps of K-means algorithm

The merge operation builds a dendrogram representing the nested clusters. The dendrogram shows the pattern and similarities of the clusters. Dendrograms can be seen as a hierarchical representation of the clusters showing the similarity or dissimilarity of them.

3.4   Artificial Neural Networks and Deep Learning

Deep learning is a type of representation learning where the machine itself learn several internal representations from raw data to perform regression or classification [23]. This is in contrast to more classical machine learning algorithms which often require carefully engineered features that are based on domain expertise [1]. Deep learning models are built up in a layer-wise structure where each layer learn a set of hidden representations, that in many cases cannot be understood by a human observer. The representations in each layer are non-linear compositions of the representations in the previous layer. This allows the model to first learn very simple representations in the first layers which are then combined into more and more complex and abstract representations for each layer. An example of this is that when deep learning models are used on images, they often start by learning to detect edges and strokes [44]. These are then combined into simple objects, objects that then are combined into even more complex objects for each layer. Since each layer only learns from the representation of the previous layer, a general purpose learning algorithm, such as back propagation [22], can be used to train a given network.

3.4.1 Artificial Neural Networks

Most algorithms in deep learning are based on artificial neural networks [23]. In contrast to deep learning the field of artificial neural networks has been around for some time. It all started in 1943 when McCulloch and Pitts, a neuroscientist and mathematician, defined a mathematical model of how they believed a neuron worked in a biological brain [28]. The next step came in 1949 when Hebb came up with a rule that made it possible to train an artificial neuron to learn and subsequently recognize a set of given patterns [16]. In 1958 Rosenblatt, a psychologist, further generalised the works of McCulloch and Pitts and proposed a model, called the *perceptron*, for an artificial neuron [39]. The mathematical definition of a perceptron is given in Eq. (3.11), and a graphical representation is shown in Fig. 3.8.

$$y = f\left(\left(\sum_i x_i * w_i\right) + b\right) \tag{3.11}$$

The perceptron was then further analysed and developed by Minsky and Papert [30]. In the analysis of the perceptron, Minsky and Papert showed that a single perceptron was not sufficient to learn certain problems (e.g., nonlinear problems), for example, the XOR problem. Instead, they argued that multi-layered perceptrons were needed to solve such problems. However such networks were not possible to train at that time, this lead to an AI winter and very little research on ANNs were conducted on neural networks for some time. This has changed during the years,

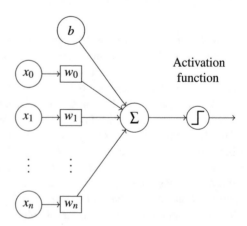

Fig. 3.8 A graphical illustration of a perceptron. The output of a perceptron is an activation function applied to the weighted sum of the inputs plus a bias. The mathematical definition of a perceptron is given by Eq. (3.11)

thanks to the increase in computational and improvements to the methodology, such as the introduction of the backpropagation algorithm, unsupervised pre-training [9] and the rectified linear unit [34]. These improvements have allowed researchers to build networks with many hidden layers, so called deep neural networks [23]. In the following sections, we present several different architectures of artificial neural networks used in deep learning.

3.4.2 Feedforward Neural Networks

A feedforward neural network is an artificial neural network where information only moves in one direction; thus feedforward networks are acyclical and therefore free of loops. The layout of a typical feedforward network is shown in Fig. 3.9. The most basic feedforward network is the perceptron [39] where the output is an activation function applied to the weighted sum of the input plus a bias. If the sigmoid function, described in Eq. (3.6), is used as the activation function a single perceptron performs exactly the same task as logistic regression (see Sect. 3.2.2). A standard architecture of feedforward networks is to arrange multiple neurons in interconnected layers. Each neuron in any layer, except the final output layer, has directed connections to all neurons in the subsequent layer. This types of networks are called *multilayer perceptrons*. As with the perceptron, the output that each neuron will propagate to the next layer is an activation function applied to the weighted sum of all inputs plus a bias. As long as the activation function is differentiable, it is possible to calculate how the output will change if any of the weights is changed, and thus the network can be optimized with gradient based methods.

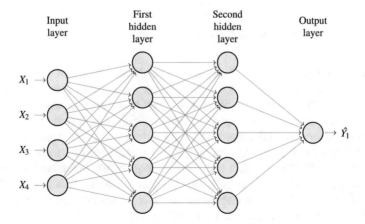

Fig. 3.9 The layout of a multilayer perceptron with two hidden layers, each having 5 neurons

3.4.3 Convolutional Neural Networks

Convolutional neural networks (CNNs) are a special type of neural networks that are mainly used in image analysis [25], but some researchers have used CNNs for natural language processing [24]. The main idea behind the CNN architecture is that basic features in a small area of an image can be analysed independently of its position and the rest of the image. Thus an image can be split up into many small patches. Each patch can then be analysed in the same way and independently of the other patches. The information from each patch can then be merged, to create a more abstract representation of the image.

This scheme is implemented in a CNN using two different steps; convolutional and sub-sampling steps. In a convolutional step, a feedforward neural network is applied to all small patches of the image, generating several maps of hidden features. In the sub-sampling step, the size of the feature map is reduced. This is often done by reducing a neighborhood of features to a single value. The most common way for this reduction is to either represent the neighborhood with the maximum or the average value. These two steps are then combined into a deep structure with several layers.

It has been shown that a CNN learns to detect general and simple patterns in the first layers, such as detecting edges, lines, and dots [44]. The abstraction of the learned features will increase in each layer. If, for example, the first layer detects edges and dots, the next layer may combine these edges and dots into simple patters. These patterns may then be combined into more complex and abstract objects in the next layer. One of the main benefits of this approach is that the CNN learns translation invariant features. Thus a CNN can learn general features about objects in an image independently of their position within the image.

3.4.4 Recurrent Neural Networks

A recurrent neural network (RNN) is a type of artificial neural network where there are cyclical connections between neurons, unlike feedforward networks which are acyclical [37]. This allows the network to keep an inner state allowing it to act on information from previous input to the network, thus exhibit dynamic temporal behaviour. This makes RNNs optimal for the analysis of sequential data, such as text [29] and time series [5]. One big problem with recurrent neural networks, which also occurs in deep feedforward networks, is that the gradients in the backpropagation will either go to zero or infinity [36]. This has however been partially solved by the introduction of special network architectures, such as the long short term memory (LSTM) [14] and the gated recurrent unit (GRU) [4].

3.4.5 Generative Adversarial Networks

Generative adversarial networks (GANs) were introduced by Ian Goodfellow in 2014 [13]. The idea behind a GAN is that we have two artificial neural networks that compete. The first network, called the generator, tries to generate examples following the same distribution as the collected data. While the second network, called the discriminator, tries to distinguish between the examples that are generated by the generator and the data that are sampled from the real data distribution.

The training of these two networks consists of two phases where the first part aims to train the generator and the second to train the discriminator. In the first phase, the generator creates several examples and gets information about how the discriminator would judge these examples, and in which direction to change these examples so that they are more likely to pass as real data to the discriminator. In the second phase, several generated and real examples are presented to the discriminator, that classifies them as real or generated. The discriminator is then given the correct answers and how to change its settings to preform better when classifying future examples. This can be compared to the competition between a money counterfeiter and a bank. The task of the counterfeiter is to generate fake money, and the bank should be able to determine if money is faked or not. If the counterfeiter gets better at creating new fake money, the bank must take new measures to discover the fake money and if the bank gets better at discovering fake money, the counterfeiter must come up with better and creative ways to create new money. The hope when training a GAN is that the generating network and the discriminating network will reach a stalemate where they are both good at their tasks. Successful works including GANs, are the generation of images of human faces [12], images of hotel rooms [38] and the generation of text tags to images [31].

3.5 Model Evaluation

There are many representations to build models from data. In the previous sections, we have seen some of them. However, we need to evaluate the built model to check how well it performs on unseen examples, that is, how well it generalizes the training dataset. The design of machine learning system, as stated before, is composed of several steps: (i) the choice of a dataset as input, (ii) the choice of a representation for a learner, (iii) an approach to optimize the model, and, finally, (iv) an evaluation of the model. The evaluation (or assessment) must be done in a dataset not used for training. Basically, the original dataset is split into two subsets: the training and the test sets. The usual approach to dividing the dataset is as follows: 70% for training and 30% for testing. Remark that the number of examples in each subset depends on the number of example in the original dataset, and the selection of examples for each subset must be balanced (mainly in classification problems), that is, each subset must have representative information of the domain to be modeled.

   Another important remark is that, during the training phase, we have to test our algorithm (or algorithms) using different hyper parameters (e.g., learning rate, node purity, number of clusters, etc.). In this scenario, we may also divide the training set into cross-validation sets (or split the original dataset into three subsets: training, validation, and test). Therefore, the training data is used to training some learning algorithms. In the validation set, the performance of trained algorithms are evaluated, and thus, the best one is chosen to model our problem. Moreover, the test data is used to evaluate the chosen model against new examples.

   In the training set, we use a loss function (or another similar function) to verify whether or not our model is converging. When we are satisfied with the results in the training set, our model is run against the test set, and, depending on the representation used, we choose a metric to evaluate the predictions made by the model. It is clear that a metric for classification is different from a metric for regression, and it is not the same for unsupervised approaches. In the following, we present some metrics for the representations discussed in the previous sections.

3.5.1   Regression

The regression learners predict continuous values, and the metrics find how close a predicted value is to the real value (the ground-truth value). Therefore, most of the approaches are based on the distance between the ground-truth and predicted values (represented by $y - \hat{y}$).

   Some popular metric functions for regression are:

- R^2 score (aka coefficient of determination) is a number that indicates the proportion of the variance in the predicted output from the real output. It is calculated as follows: $R^2 = 1 - \frac{\sum_{i=1}^{m}(y^{(i)}-\hat{y}^{(i)})^2}{\sum_{i=1}^{m}(y^{(i)}-\bar{y}^{(i)})^2}$, where m is the size of the (test) dataset, $\hat{y}$ is the predict value, and $\bar{y}$ is the average of ground-truth values. The closer R^2 is to one, the better.
- Mean square error (MSE) measures the average of the squares of the errors of the predict and real value. The error means the difference between the two values (predicted and truth). The differences are also called residuals. The mathematical definition is given as follows: $MSE = \frac{1}{m}\sum_{i=1}^{m}(y^{(i)} - \hat{y}^{(i)})^2$. The closer MSE is to zero, the better.
- Root mean square error (RMSE) is a measure of the differences between values (sample and population values) predicted by a model or an estimator and the truth-ground values. RMSE is the square root of the value calculated by MSE. So, its mathematical definition is $RMSE = \sqrt{2MSE}$. The closer RMSE is to zero, the better.

- Mean absolute error (MAE) is similar to MSE, but it does not square the error. The absolute value of the difference is used instead. It is defined as follows: $MSE = \frac{1}{m} \sum_{i=1}^{m} |y^{(i)} - \hat{y}^{(i)}|$.
- Mean absolute percent error (MAPE) is another metric to evaluate regression models, and the error expressed in generic percentage terms is: $MAPE = (\frac{1}{n} \sum_{i=1}^{m} \frac{|y^{(i)} - \hat{y}^{(i)}|}{|y^{(i)}|}) \times 100$

MAPE and MAE are less sensitive to the occasionally very large error because they do not square the errors. Therefore, if we want our model to ignore big prediction errors, MAPE and MAE may be used. However, the metric which is considered as *one size fits all* is RMSE [43].

3.5.2 Classification

The metrics for classification problems are a little bit easier to apply on the model than the regression ones. Roughly speaking, the metrics are based on counting how many predicted classes equals to the observed ones. We use the binary classification to present the metrics since it is more intuitive to understand. The same reason is applied to multi-class prediction.

The simpler way to evaluate a classifier is when the classes are well balanced in the dataset (training and test). Accuracy is the metric for this scenario. The predicted classes are matched against to the observed ones, and the number of matched ones is divided by the size of the dataset: $\frac{1}{m} \sum_{i=1}^{m} y^{(i)} == \hat{y}^{(i)}$, considering that *false* is 0, and *true* is 1.

In most of the cases, the classes in a dataset are skewed, that is, the number of classes is not balanced. For instance, in a dataset with examples of benign and malignant tumors, maybe most of the examples are labeled as benign tumors. If 96% of the tumors are labeled as benign, and the model outputs benign for every example, its accuracy will be 96% (a very good accuracy); however, we know that the model is not able to predict malignant tumors.

To overcome this problem, several metrics have been proposed, and we discuss three of them: precision, recall, and F_1-score.[5] First, we present some definitions to help present the metrics above. Table 3.5 presents a taxonomy of class classification that is used to classify the predicted classes in relation to the ground-truth ones. The true positive (TP) means that the classifier matches the positive classes, and the false positive (FP) implies that the classifier outputs negative classes as positive ones. The same reasoning for negative classes: true negative (TN) indicates negative classes are predicted correctly, and false negative (FN) indicates positive classes are predicted as negative ones. We remark that the confusion matrix can be extended for the multi-class problem [2].

---

[5]F_1-score is a specialization of F_β-score that is not covered in this chapter.

Table 3.5 The confusion matrix

	Positive	Negative
Positive	True positive (TP)	False positive (FP)
Negative	False negative (FN)	True negative (TN)

Precision, recall, and F_1-score can be defined as follows:

$$precision = \frac{TP}{TP + FP}$$

$$recall = \frac{TP}{TP + FN}$$

$$F_1\text{-}score = \frac{2 \times precision \times recall}{precision + recall}$$

The precision metric is used when we want exactness, that is, our classifier is covering the positive classes confidently. On the other hand, recall means completeness, that is, how many positive examples our classifier has missed. If we want to balance between recall and precision, F_1-score gives the harmonic mean of precision and recall.

Remark that accuracy can be also calculated as follows: $\frac{TP + TN}{TP + FP + TN + FN}$.

3.5.3 Clustering

The evaluation of the clusters resulted from a clustering algorithm is a not easy task since there are no true labels to compare with the clusters. The evaluation can be divided into two categories: internal and external. The internal category measures the quality using the training data, and the external category uses the external data (test set). However, the external evaluation is not completely accurate as compared to the methods for supervised learning [10].

Silhouette Coefficient is a (internal) metric to measure the quality of the built clusters. It is a popular method that combines both cohesion (similarity between an object and its cluster) and separation (similarity between an object and other clusters). The silhouette coefficient for an individual object (example or data point) can be computed as follows: (i) given an example e_i, calculate its average distance to the other examples in the same cluster (a_{ei}), (ii) do the same using the other clusters (b_{ei}), and (iii) the silhouette coefficient for e_i is $s_i = \frac{b_i - a_i}{max(b_i, a_i)}$. The average of all coefficients s_k can be calculated to find the clustering coefficient.

Remark that the coefficient can be a value between -1 and 1. A negative value of s_i is not desirable since it indicates that the average distance of e_i to its cluster is greater than to the other clusters. On the other hand, a positive value of s_i is an ideal value, and $s_i = 1$ indicates that the the the average distance of a_{ei} is 0.

Rand Index (RI) is a metric for external evaluation. It compares the predicted clusters to the real clusters (manually assigned by an expert user), and it is similar to the accuracy metric for supervised algorithms. Here is how to calculate RI:

1. Let X be a dataset, C_p be a clustering (set of clusters) build by the clustering algorithm (predicted), and C_r be a set of ground-truth clusters;
2. TP is the number of examples (data points) belonging to the same clusters in C_p and C_r;
3. TN is the number of examples (data points) belonging to different clusters in C_p and C_r;
4. FP is the number of examples (data points) belonging to a cluster in C_p but to a different cluster in C_r;
5. FN is the number of examples (data points) belonging to a different cluster in C_p but the same cluster in C_r
6. Rand Index is calculated as follows: $RI = \frac{TP+TN}{TP+TN+FP+FN}$.

We remark that RI is calculated exactly as accuracy.

Metrics are essential tools to evaluate a machine learning system. Each one must be carefully studied to understand the behavior of the built model. Some metrics can be affected by noise in the data (aka outliers), and others may smooth the effects of noise.

3.6 Dimensionality Reduction

Dimensionality reduction plays an essential role in machine learning. Its goal is to decrease the number of features of a dataset. As an example, let's suppose that we want to identify objects in a set of images. Each image is a 100×100 pixels, thus we have $10,000$ features. An approach to reduce the dimensionality can bring the number of feature to 1000. Therefore, dimensionality reduction may be applied to:

- Compress data in the main and secondary storage.
- Speed up learning algorithms.
- Visualize the dataset in 2D or 3D planes.

The reduction may also merge the more correlated features into one (or more). For instance, a dataset can have a feature f_1 that represents the height in centimeters and another feature f_2 that also represents the height but in inches. Based on the correlation of f_1 and f_2, a dimensionality reduction technique may merge both features to form a new one.

Given a dataset $X = \{(x^{(1)}, y^{(1)}), \ldots, (x^{(m)}, y^{(m)})\}$ (possibly without labels $y^{(i)}$), the dimensionality reduction aims to transform $x^{(i)} \in \mathbb{R}^d$ into $z^{(i)} \in \mathbb{R}^k$ (where $k < d$), resulting in $X' = \{(z^{(1)}, y^{(1)}), \ldots, (z^{(m)}, y^{(m)})\}$. Figure 3.10 shows a 3D dataset

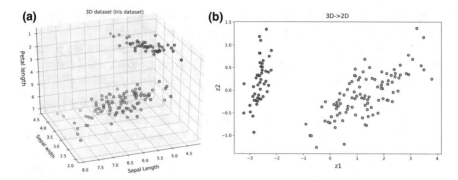

Fig. 3.10 A 3D dataset reduced to 2D dataset

(three features a) reduced to a 2D dataset (two features b). The reduction was done using the PCA technique.

3.6.1 Principal Component Analysis (PCA)

PCA reduces the dimensionality of a dataset by projecting vectors onto the plane and minimizing the projection distance error between the points and the projected vector. It can be described as follows: given a dataset with d dimensions, find k vectors $\mu^{(1)}$, ..., $\mu^{(k)}$ onto which to project the data, so as to minimize the projection squared error.

To find the vectors the following steps must be done: (i) find the covariance matrix of the dataset: $\Sigma = \frac{1}{m} \sum_{i=1}^{m} x^{(i)} \cdot x^{(i)^T}$ (Σ is a $d \times d$ matrix), calculate the eigenvector[6] U of Σ (U is a $d \times d$ matrix), and reduce X to k-dimensions based on U as follows $X_{reduced} = U[:, 1:k]^T \cdot X^T$. Remark that the new number of dimensions is taken from U which represents the vectors $\mu(i)$. We can approximately reconstruct X from $X_{reduced}$ and $U[:, 1:k]$ as follows $X_{app} = U[:, 1:k] \cdot X_{reduced}$

The ideal number of new dimensions (k) is found giving the percentage of variance to retain. If we want to retain 99% of variance, the threshold φ must be set to 0.01. We find the threshold as follows: $\frac{\frac{1}{m}\sum_{i=1}^{m}|x^{(i)}-x_{app}^{(i)}|^2}{\frac{1}{m}\sum_{i=1}^{m}|x^{(i)}|} \leq \varphi$. To find the best k, we test the variance retained with $k = 1, \ldots, d-1$ until we reach the desired threshold.

PCA performs a linear mapping to merge points in the plane. However, some features in a dataset may have more complex polynomial relationships. Therefore, there are some techniques to deal with nonlinear dimensionality reduction. t-Distributed Stochastic Neighbor Embedding (t-SNE) is an approach based on probability distributions with a random walk on neighborhood graphs to find the structure within the data [27]. t-SNE converts distances between data in the original space to probabilities.

---

[6]Definition of eigenvectors can be found in traditional books of linear algebra.

Another approach that works with nonlinear dimensionality is an extension of PCA: nonlinear PCA. Nonlinear PCA is performed using a five-layer neural network [35] that captures the complex nonlinear relationship between the features.

The decision in using a linear or nonlinear approach must be based on the characteristic of the dataset; however, sometimes it is not easy to identify whether or not features are linearly or nonlinearly correlated. A rule of thumb is to start with PCA. If it does not work well, use more sophisticated techniques.

3.7   Final Remarks

The design of a machine learning system is composed of several steps: from choosing a domain (dataset) to building a model to accurately predict new information about the given domain. This section closes this chapter by presenting some issues that are orthogonal to the subjects discussed so far. An essential step in machine learning is to prepare the dataset as a proper input of a learning algorithm. The quality of the dataset has a high impact on the performance of machine learning-based methods. Thus, in this section, we present some issues about preprocessing the data and checking the behavior of the learning algorithm.

3.7.1   Data Preprocessing

Some characteristic of a dataset may have adverse influence on the learning results or even may be not suitable for certain learning algorithm classes. Several learning algorithms deal with only numerical values, and, in this case, features that are not numerical must be either discarded or transformed into numerical values (discretization may be applied). For example, a feature that stores gender as m and f may have the values replaced by the numerical values 1 and 2, respectively. The contrary is also true: a numerical value can be transformed into a string to speed up some learning algorithms (e.g., decision trees).

A dataset can also have features on very different scales, that is, one feature can store values on the order of thousands, and another on the order of ten. An example is the price of an apartment and number of bathrooms. In this case, we can use a technique called feature scaling, that is, all features are scaled to a same range of values. The two most common technique are: standardization ($\frac{X - \bar{X}}{\sigma}$) and rescaling ($\frac{X - min(X)}{max(X) - min(X)}$) feature rescaling. The former makes all features to have the average close to 0, and the latter the features will have values between 0 and 1. The feature scaling helps an optimizer to converge faster.

Missing data is another problem that must be solved during the preprocessing stage.[7] Three approaches may be used to address it: analyze only the available data, impute missing values in the dataset, and use a learning algorithm that deals with missing data.

For the first case, examples or features are deleted from the dataset. This is recommended when there are not many missing values. The dataset size allows the removal of some examples or features, and the values are randomly missing.

Missing values' imputation aims to replace missing values with some plausible values. The new values are calculated based on some traditional statistic methods (e.g., mean, the most frequent value, or median), or some other more sophisticated approaches (e.g., Expectation Maximization [7], Shell Neighbor Imputation [45]).

The third approach is to use a learning algorithm that integrates components to deal with missing values. Probably, these extensions of traditional learning algorithms use some statistical methods cited previously, e.g., [41].

In the dataset preprocessing step, we can also drop unnecessary features (aka attribute or feature selection) or create new features based on the existing ones (aka attribute or feature transformation). Imagine a dataset with characteristics of cars, and we want to learn the safety of the cars (e.g., low, medium, high). A feature like license plate will be not important for the learning algorithm. On the other hand, we can build new features from existing ones to improve the learning algorithm performance, e.g., based on the *weight* of the car and its *horsepower*, a new feature *power-to-weight ratio* can be created. The feature creation is very useful, for example, when we want to build a polynomial model for linear regression (as we saw in Sect. 3.2.1.1).

Future selection is a largely used tool in preprocessing dataset to improve a learning algorithm's performance, and there are many approaches to accomplish it. Most of them identify the relevance of a feature in relationship to the others. Based on the relevance, a subset of features can be extracted from the original dataset [33].

The preprocessing step plays an essential role to build good machine learning systems. There is no good rule of thumb to guide during this step. However, the best thing to do is to test several approaches by assessing the results. Another issue to consider is that an approach may fit very well in one dataset but may have poor performance in another one.

3.7.2   Bias, Variance, Under, and Overfitting

During the learning step, the model may suffer from some learning problems. The most two common ones are under and overfitting. They are intimately related to the bias and the variance of the model. Bias and variance are used to identify some issues in assessing the ability of a learning method to generalize.

---

[7]For the sake of simplicity, we consider an invalid value (e.g., mixed characters and numerical values) for a feature as missing data too.

Roughly speaking, variance means how much structure from the dataset the model has learned, while bias means how much structure from the dataset the model has not learned. That is, bias is a learner tendency to learn the same wrong thing consistently, and variance is the tendency to learn random things regardless the dataset [8]. Intuitively, a biased model has a poor performance in the training set and in the test set (as expected) while, a model with variance has good performance in the training set but poor performance in the test set. When a model has high variance, it means that the model has caught all the details of the training set (including noise and outliers), thus cannot be generalized to unseen examples.

Bias and variance can be identified by verifying the performance of the model in both the training and test set. Considering that a model performs better in the training set, the behavior in the test set must follow the performance in the test sets. The ideal scenario would be with low bias and low variance, that is, neither does the model make a strong assumption regarding the dataset nor does it learn useless characteristics from the dataset.

The overfitting and underfitting problems are usually fixed using a (cross) validation dataset. Models and their parameters are trained in the training set, ranked in the validation set, and the best model is evaluated in the test set. Another tool widely used to combat the over/underfitting is to add a regularization term to the evaluation function. For example, we can add to the gradient descent equation (Eq. 3.4) a regularization term:

$$\Theta = \Theta - \alpha [\frac{1}{m} \sum_{i=1}^{m} (h_\Theta(x^{(i)}) - y^{(i)})x^{(i)} + \lambda \sum_{j=1}^{m} \Theta_j] \qquad (3.12)$$

where λ represents the strength of the regularization. Clearly, if λ is equal to 0 then there is no regularization (or penalty).

Multi-label and multi-target[8] algorithms are classifiers that learn a vector of values from the observed data (examples). Most traditional supervised learning algorithms are extended to deal with multi-label or multi-target values. We do not cover these extensions in this chapter, and details can be found in [17].

3.8 Conclusions

In this chapter, we presented a comprehensive overview of various aspects of machine learning. The subjects introduced are widely discussed in the literature; however, this chapter intends to be a starting point for those who are interested in applying machine learning techniques to the real-world problems.

---

[8] Also known as multivariate or multi-output regression.

Even though, we find many researches on machine learning, it still remains a young field with many under-explored research opportunities. In addition, it has a lot of *folk wisdom* that can be hard to come by, but helps its development [8, 20].

Besides, machine learning plays an important role for data science. The knowledge brought by machine learning in building models for prediction makes it an essential tool for those wanting to extract information and knowledge from data.

Acknowledgements Denio Duarte is partially funded by Coordenadoria de Aperfeiçoamento de Pessoal de Nível Superior (CAPES) under process number 88881.119081/2016-01—Brazil during his visit to Skövde Artificial Intelligence Laboratory (SAIL) at University of Skövde (HiS).

References

1. Bengio, Y., Courville, A., & Vincent, P. (2013). Representation learning: A review and new perspectives. *IEEE Transactions on Pattern Analysis and Machine Intelligence, 35*(8), 1798–1828.
2. Boutell, M. R., Luo, J., Shen, X., & Brown, C. M. (2004). Learning multi-label scene classification. *Pattern Recognition, 37*(9), 1757–1771.
3. Bratko, I., Michalski, R. S., & Kubat, M. (1999). *Machine learning and data mining: Methods and applications.*
4. Chung, J., Gulcehre, C., Cho, K., & Bengio, Y. (2015). Gated feedback recurrent neural networks. In *International Conference on Machine Learning* (pp. 2067–2075).
5. Connor, J. T., Martin, R. D., & Atlas, L. E. (1994). Recurrent neural networks and robust time series prediction. *IEEE Transactions on Neural Networks, 5*(2), 240–254.
6. De Houwer, J., Barnes-Holmes, D., & Moors, A. (2013). What is learning? On the nature and merits of a functional definition of learning. *Psychonomic Bulletin & Review, 20*(4), 631–642.
7. Dempster, A. P., Laird, N. M., & Rubin, D. B. (1977) Maximum likelihood from incomplete data via the EM algorithm. *Journal of the Royal Statistical Society. Series B (Methodological),* 1–38.
8. Domingos, P. (2012). A few useful things to know about machine learning. *Communications of the ACM, 55*(10), 78–87.
9. Erhan, D., Bengio, Y., Courville, A., Manzagol, P. A., Vincent, P., & Bengio, S. (2010). Why does unsupervised pre-training help deep learning? *Journal of Machine Learning Research 11,* 625–660.
10. Färber, I., Günnemann, S., Kriegel, H. P., Kröger, P., Müller, E., & Schubert, E., et al. (2010). On using class-labels in evaluation of clusterings. In *Multiclust: 1st International Workshop on Discovering, Summarizing and Using Multiple Clusterings Held in Conjunction with KDD* (p. 1)
11. Fujita, A., Takahashi, D. Y., & Patriota, A. G. (2014). A non-parametric method to estimate the number of clusters. *Computational Statistics & Data Analysis, 73,* 27–39.
12. Gauthier, J. (2014). Conditional generative adversarial nets for convolutional face generation. In *Class Project for stanford CS231N: Convolutional neural networks for visual recognition* (Vol. 2014, No. 5, p. 2). Winter Semester
13. Goodfellow, I., Pouget-Abadie, J., Mirza, M., Xu, B., Warde-Farley, D., & Ozair, S., et al. (2014). Generative adversarial nets. In *Advances in neural information processing systems* (pp. 2672–2680).
14. Graves, A. (2013). *Generating sequences with recurrent neural networks.* arXiv:1308.0850.

15. Gunn, S. R. (1998). Support vector machines for classification and regression. Technical report, Faculty of Engineering, Science and Mathematics–School of Electronics and Computer Science.
16. Hebb, D. (1949). *The organization of behavior: A neuropsychological theory*. Wiley
17. Izenman, A. J. (2008). Modern multivariate statistical techniques. Regression, classification and manifold learning.
18. Jain, A. K. (2010). Data clustering: 50 years beyond k-means. *Pattern Recognition Letters, 31*(8), 651–666.
19. Jain, A. K., Murty, M. N., & Flynn, P. J. (1999). Data clustering: A review. *ACM Computing Surveys (CSUR), 31*(3), 264–323.
20. Jordan, M. I., & Mitchell, T. M. (2015). Machine learning: Trends, perspectives, and prospects. *Science, 349*(6245), 255–260.
21. Keogh, E., & Mueen, A. (2010). *Curse of dimensionality*. US: Springer.
22. Le Cun, Y., Touresky, D., Hinton, G., & Sejnowski, T. (1988). A theoretical framework for back-propagation. In *The connectionist models summer school* (Vol. 1, pp. 21–28).
23. LeCun, Y., Bengio, Y., & Hinton, G. (2015). Deep learning. *Nature, 521*(7553), 436–444.
24. LeCun, Y., Bottou, L., Bengio, Y., & Haffner, P. (1998). Gradient-based learning applied to document recognition. *Proceedings of the IEEE, 86*(11), 2278–2324.
25. LeCun, Y., Jackel, L., Bottou, L., Brunot, A., Cortes, C., & Denker, J., et al. (1995). Comparison of learning algorithms for handwritten digit recognition. In *International Conference on Artificial Neural Networks*, Perth, Australia (Vol. 60, pp. 53–60).
26. Loh, W. Y. (2011). Classification and regression trees. In *Wiley interdisciplinary reviews: Data mining and knowledge discovery* (Vol. 1, No. 1, pp. 14–23).
27. Maaten, L. V. D., & Hinton, G. (2008). Visualizing data using t-SNE. *Journal of Machine Learning Research 9*, 2579–2605.
28. McCulloch, W. S., & Pitts, W. (1943). A logical calculus of the ideas immanent in nervous activity. *The Bulletin of Mathematical Biophysics, 5*(4), 115–133.
29. Mikolov, T., & Zweig, G. (2012). Context dependent recurrent neural network language model. *SLT, 12*, 234–239.
30. Minsky, M., & Papert, S. (1969). *Perceptrons*.
31. Mirza, M., & Osindero, S. (2014). *Conditional generative adversarial nets*. arXiv:1411.1784.
32. Mitchell, T. M. (1997). *Machine learning* (1st ed.). New York, NY, USA: McGraw-Hill Inc.
33. Molina, L. C., Belanche, L., & Nebot, A. (2002). Feature selection algorithms: A survey and experimental evaluation. In *Proceedings of the 2002 IEEE International Conference on Data Mining* (pp. 306–313).
34. Nair, V., & Hinton, G. E. (2010). Rectified linear units improve restricted boltzmann machines. In *Proceedings of the 27th International Conference on Machine Learning* (ICML 2010) (pp. 807–814).
35. Oja, E. (1997). The nonlinear PCA learning rule in independent component analysis. *Neurocomputing, 17*(1), 25–45.
36. Pascanu, R., Mikolov, T., & Bengio, Y. (2013). On the difficulty of training recurrent neural networks. *ICML, 3*(28), 1310–1318.
37. Pineda, F. J. (1987). Generalization of back-propagation to recurrent neural networks. *Physical Review Letters, 59*(19), 2229.
38. Radford, A., Metz, L., & Chintala, S. (2015). *Unsupervised representation learning with deep convolutional generative adversarial networks*. arXiv:1511.06434.
39. Rosenblatt, F. (1958). The perceptron: A probabilistic model for information storage and organization in the brain. *Psychological Review, 65*(6), 386.
40. Samuel, A. L. (1959). Some studies in machine learning using the game of checkers. *IBM Journal of Research and Development, 3*(3), 210–229.
41. Smola, A. J., Vishwanathan, S. V. N., & Hofmann, T. (2005). Kernel methods for missing variables. In R. G. Cowell, & Z. Ghahramani (eds.) *Proceedings of the Tenth International Workshop on Artificial Intelligence and Statistics* (pp. 325–332). Society for Artificial Intelligence and Statistics.

42. Weisberg, S. (2005). *Applied linear regression* (Vol. 528). Wiley
43. Willmott, C. J. (1982). Some comments on the evaluation of model performance. *Bulletin of the American Meteorological Society*, *63*(11), 1309–1313.
44. Zeiler, M. D., & Fergus, R. (2014) Visualizing and understanding convolutional networks. In *European Conference on Computer Vision* (pp. 818–833). Springer.
45. Zhang, S. (2011). Shell-neighbor method and its application in missing data imputation. *Applied Intelligence*, *35*(1), 123–133.

Part II
Application Domains

Chapter 4
Information Fusion

H. Joe Steinhauer and Alexander Karlsson

Abstract The study of information fusion comprises methods and techniques to automatically or semi-automatically combine information stemming from homogeneous or heterogeneous sources into a representation that supports a human user's situation awareness for the purposes of decision making. Information fusion is not an end in itself but studies, adapts, applies and combines methods, techniques and algorithms provided by many other research areas, such as artificial intelligence, data mining, machine learning and optimization, in order to customize solutions for specific tasks. There are many different models for information fusion that describe the overall process as tasks building upon each other on different levels of abstraction. Information fusion includes the analysis of information, the inference of new information and the evaluation of uncertainty within the information. Hence, uncertainty management plays a vital role within the information fusion process. Uncertainty can be expressed by probability theory or, in the form of non-specificity and discord, by, for example, evidence theory.

4.1 Introduction

A typical task for a data scientist is to make new discoveries form data. This data can stem from one or several sources; it can be homogeneous or heterogeneous, represent a snapshot of a situation, a time series, etc. The data might include inconsistencies or might be uncertain, e.g. is probabilistic or fuzzy. The study of information fusion comprises methods and techniques to automatically or semi-automatically combine information stemming from homogeneous or heterogeneous sources into a representation that supports a human user's understanding of the observed situation. The general idea behind this is that better informed decisions can be made on the basis of more and better information. Hence, information fusion includes the analysis

H. J. Steinhauer (✉) · A. Karlsson
University of Skövde, Skövde, Sweden
e-mail: joe.steinhauer@his.se

A. Karlsson
e-mail: alexander.karlsson@his.se

© Springer International Publishing AG, part of Springer Nature 2019
A. Said and V. Torra (eds.), *Data Science in Practice*, Studies in Big Data 46,
https://doi.org/10.1007/978-3-319-97556-6_4

of information, the inference of new information and the evaluation of uncertainty within the information. Information fusion is not an end in itself but studies, adapts, applies and combines methods, techniques and algorithms provided by many other research areas, such as artificial intelligence, data mining, machine learning and optimization, in order to customize solutions for specific tasks. It has many different application areas, such as computer vision, forensics, biometrics, robotics, network security, diagnosis, surveillance, smart cities, autonomous vehicles and networked mobile devices.

An example of information fusion occurring in nature is a human keeping its balance while walking. Here, the sensory input from vision (sight), proprioception (touch) and the vestibular system (motion, equilibrium, spatial orientation) are integrated together with previously learned information (e.g. that icy roads are slippery) before the brain will send the information to the various body parts about how to make adjustments of the body's position [1].

In comparison, an autonomous vehicle needs to integrate information from the camera with that it follows the lanes on the street, the GPS system with that it tracks its overall position, a thermometer that measures the temperature of the road, several proximity sensors giving information how close by the nearest objects are, the speedometer and several more devices in order to be able to drive safely through urban territory.

A system that applies information fusion might do this on many different levels of abstraction and for many different tasks. Consequently, the purposes of information fusion are depended on the respective tasks, but they can often be identified to be one of the following:

1. To increase the dimensionality of available information by inferring new information when previously obtained information is brought together.

2. To increase accuracy in information by utilizing numerous information sources.

3. To decrease uncertainty in order to provide better grounds for decision making.

4. To decrease dimensionality of relevant data in order to provide a, for a human user, cognitively easy to comprehend abstraction of a situation. This effort is closely related to visual data analysis that is described in [4] (a chapter this book).

This chapter presents a short introduction to the area of information fusion. As the area is complex and comprises many different aspects we cannot provide a complete overview within the scope of this book but have chosen some of the major aspects of information fusion. In Sect. 4.2 several models that illustrate the information fusion process and its different stages are explained. As information fusion is often used within decision support systems where one challenge is to improve a human user's situation understanding, the achievement of situation awareness, that is described in Sect. 4.3, goes hand in hand with the information fusion task. Many methods used for information fusion originate from the area of machine learning such as, clustering and classification which are closer described in [10] (a chapter in this book). They are frequently used by data scientists for data mining purposes which, in turn can aid the information fusion process and vice versa, as described in Sec. 4.4. A typical task in information fusion is to fuse information stemming from different sources in order to describe or predict the state of a situation or object. The information form each source can hence be regarded as evidence towards a certain state. How evidence can be expressed and combined is therefore described in Sect. 4.5. Information might be uncertain, hence, to be able to express, handle and measure uncertainty is a relevant aspect of information fusion and is discussed in Sect. 4.6.

4.2   Models for Information Fusion

The Joint Directory of Laboratories (JDL) (e.g. [13]) has developed a model for information fusion, shown in Fig. 4.1, that is often used to illustrate the different tasks within the information fusion system. The model divides the information fusion process into five different types of tasks, each taking place at an assigned abstract level within the process. These levels are:

Level 0: Signal assessment: Raw data stemming from sensors is preprocessed at the sub-object level, e.g. at pixel or signal level. Signal processing, bias corrections and unit conversion are situated at this level.

Level 1: Object assessment: Data is combined to identify objects and their features, e.g. their location, track, identity, and type.

Level 2: Situation assessment: Relationships between identified objects are inferred.

Level 3: Impact assessment: Based on the identified situation on Level 2, it is predicted how the situation might develop in the future.

Level 4: Process refinement: Resource management for the ongoing information fusion process itself, including what sources are to be used and how they are to be configured.

As information fusion processes are often used to empower human decision makers to make better informed decisions, a fifth level, explicitly dealing with user interaction, was proposed by [14].

Level 5: User refinement: Improves user interaction, e.g. by considering specific user needs and adapting to individual users.

Often, a distinction is made between *low level* and *hight level* information fusion. Low level information fusion (LLIF) subsumes the tasks situated at level 0 and

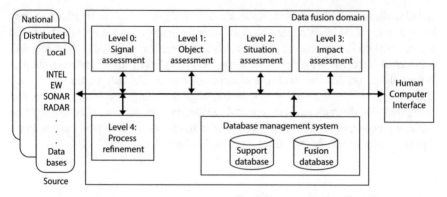

Fig. 4.1 The JDL model of information fusion (adapted from Steinberg et al. [28])

level 1 in the JDL model. This includes dealing with numerical data and concerns source processing, identification of objects, their locations and their tracks (e.g. when observing moving vehicles). High level information fusion (HLIF), on the other hand, subsumes the tasks situated at level 2 and level 3; hence, it deals with symbolic information and is concerned with how the information, detected by low level information fusion, can be interpreted and what impact they have (i.e. how they influence the situation) [31].

The purpose of information fusion is often to provide information in a way that a human user can gain a good awareness of the ongoing situation in the real word. The user should be able to interact with the system not only at the end of the information fusion process but on all levels of it during the ongoing process. This is represented in the User Fusion Model [6] shown in Fig. 4.2. According to this model, the user can influence the information fusion process after their own needs depending on

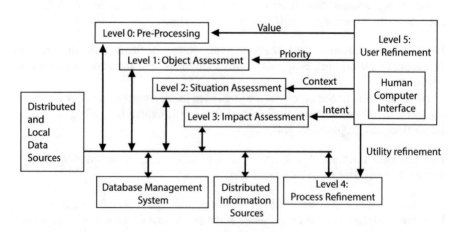

Fig. 4.2 The User Fusion Model model (adapted from Blasch [6])

the circumstances in any situation present. Thereby, the users' experience, context knowledge and their specific requirements of information in the moment can be taken into account.

Many other models for information fusion and the related area of situation awareness, which will be addressed in the next section, can be found in the literature, e.g. in [5, 6].

4.3 Situation Awareness

A person who needs to act timely and appropriately in a given situation needs to have the best possible understanding of the situation, the actions possible in this situation and their consequences. To provide the means for this awareness is, therefore, a vital task within a semi-automated system. Human situation awareness, as described further in [11], consists of:

1. Perceiving the relevant elements and their attributes in the environment.

2. Comprehending the situation by understanding in what way the elements and their relationships are of importance.

3. Being able to project the situation into the relevant future and being able to anticipate what will happen when certain actions were to be applied.

Human decision making is said to be based on the human's mental model of a situation, which is the human' s internal representation of the situation. It is needless to say, that the better this model represents the crucial elements of the actual situation, the better grounds the human has for decision making. A model often used to illustrate the human decision making process is the extended Observe-Orient-Decide-Act (OODA) loop [7] presented in Fig. 4.3. According to this model human decision making consists of:

Observe: During the observe phase information is taken in.

Orient: During the orient phase it is understood what the observed information means in the context of the situation. Newly observed information is integrated with information that has been observed during previous observe phases and with other internally represented knowledge.

Decide: During the decide phase it is determined which of the available actions is most appropriate in the situation.

Act: During the act phase the chosen action is executed.

Situation awareness and information fusion go hand in hand, hence, the four phases of the OODA loop can be matched on the information fusion process, e.g. in

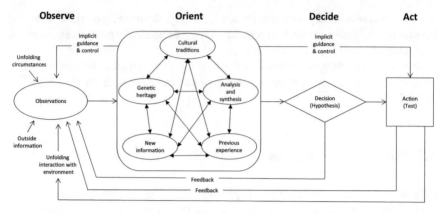

Fig. 4.3 The OODA loop (adapted from Boyd [7])

the JDL model including level 5 for user refinement (or the DFIG model that can be found in e.g. [5, 6].). Level 1 fusion can be regarded as Observe, where objects are assessed; level 2 fusion can be seen as Orient where the situation is assessed. Level 5 fusion incorporates Decide, and finally, level 4 fusion contains Act [5].

4.4 Information Fusion Versus Data Mining

As previously mentioned, information fusion is not an end in itself, but uses methods and techniques from several other areas, such as artificial intelligence, data mining and machine learning, depending on what kind of information needs to be fused. The area of data mining and information fusion show some similarities, but they also complement each other [30]. The main difference is that data mining takes data as input and outputs a model or a pattern that describes the data [15], whereas information fusion combines data and/or information to estimate and/or predict the states of entities [13].

Traditionally, data mining is an off-line process that deals with large amounts of data using batch processing. The process is inductive and the results describe properties of the data in form of models or patterns, e.g. in the form of rules or clusters. Information fusion is instead traditionally, an on-line and often real time process that usually works on relatively small sets of data using sequential processing to achieve a deductive and momentary detection, estimation or classification of the situation. It can be seen as a representation of a snapshot of the environment.

Information fusion and data mining complement each other in two ways: Firstly, information fusion can be used to pre-process data before the application of data mining techniques. It can, for example, be used to reduce the error and/or the uncertainty in data, or to increase the dimensionality of the data. It can also be applied after the data mining process to fuse different models that have been established

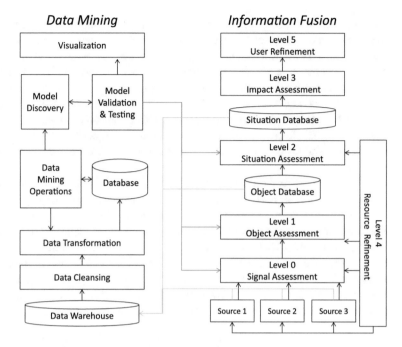

Fig. 4.4 Data mining and information fusion (based on [30])

by different data mining techniques, into one model. Secondly, data mining can be used to extract information from the data that then builds the input for the information fusion process. The latter can take place for different levels of the information fusion process, as predictive and descriptive models can potentially be used for data filtering (level 0), object recognition (level 1), situation recognition (level 2), and impact assessment (level 3). Figure 4.4 illustrates how data mining and data fusion can support each other.

4.5 Evidence

A central part of information fusion is the combination of information from different sources in order to identify which of several possible alternatives (e.g. situations or objects) is currently observed. The *frame of discernment* represents the predefined set of alternatives to chose from. Evidence combination starts with collecting the evidence and expressing it using a, for the task suitable, *uncertainty management method* (*UMM*). In this chapter, probability theory and evidence theory are considered, while others, as listed above and in [29] (a chapter in this book), can be found in the literature.

The information stemming from each source itself can be interpreted on its own and can be said to provide evidence to one or more of the alternatives. Consider, for example, a traffic surveillance system that must determine if the observed object is a pedestrian, a cyclist, or a car. A source that can roughly obtain the length of the object, might consider it to be a cyclist or a small car. A second source, estimating the width of the object, would suggest it is either a pedestrian or a cyclist. A third source might also regard it to be a pedestrian or cyclist, based on the object's position on the road. Lastly, a fourth information source would, due to the object's slow movement, emphasize that it should be a pedestrian, but that it could also be a slow moving bicycle or car.

Exactly how the observed information is translated into a probability distribution (for probability theory) or mass distribution (for evidence theory) is dependent on the task at hand, and is often based on the knowledge of human experts and/or statistics. The main difference between probability theory and evidence theory is that in probability theory the probability mass must be distributed between the singletons within the frame of discernment whereas in evidence theory it is also possible to assign a probability mass to sets of singletons. For example, when tossing a coin, there are two scenarios: either it is known whether the coin is fair or not, or there is nothing known about the coin. For both cases the frame of discernment is:

$$\Omega_X = \{heads, tails\}$$

Here, X denotes the variable in question, in this case the coin that can take the state of either $x_1 = heads$, or $x_2 = tails$. The frame of discernment Ω_X therefore can also be called a *state space*. The most appropriate probability distribution utilizing probability theory in both cases, when it either is known that the coin is fair, but also when nothing about the coin's fairness is known, would be:

$$p(heads) = 0.5$$
$$p(tails) = 0.5$$

In the first case this is intuitively correct as it is known that each alternative has a probability of 50%. However, in the second case nothing is known about the coin's fairness, and hence, nothing is known about the probability distribution. In probability theory this case is handled in the way that all probability is equally divided between all possible alternatives. On the contrary, utilizing evidence theory, it can be distinguished between these two cases. In case it is known that the coin is fair the mass function would be:

$$m(\{heads\}) = 0.5$$
$$m(\{tails\}) = 0.5$$
$$m(\{heads, tails\} = 0.0$$

but in the case where nothing about the coin's fairness is known the mass function becomes:

$$m(\{heads\}) = 0.0$$
$$m(\{tails\}) = 0.0$$
$$m(\{head, tails\}) = 1.0$$

The latter expresses that, as there is nothing known about the coin's fairness, there is no reason to particularly believe that the coin will land heads up, therefore $m(\{heads\}) = 0$; neither is there any reason to particularly believe it will land tails up, hence $m(\{tails\}) = 0$. However, it is known that the coin will land on one side or the other, which are the only alternatives that the frame of discernment allows for, hence $m(\{head, tails\}) = 1$. This expresses ambiguity in the form of non-specificity. It also means that it is possible in evidence theory to explicitly express "I don't know" by assigning all mass to the complete frame of discernment.

In the following example, regarding crime investigation, assigning evidence mass to subsets of the frame of discernment appears to be more intuitive than to establish a probability distribution from what is known. Let the frame of discernment denote three potential suspects as:

$$\Omega_X = \{Mary, John, Carol\}$$

Assume that both, Mary and John, have each an individual motive for the crime. A witness report states that a female person was seen at the crime scene. A second witness report states that a person with dark hair was seen at the crime scene, and we know that both Mary and John have dark hair whereas Carol is blond. These pieces of evidence can be transformed into a mass distribution e.g. as:

$$m(\{Mary\}) = 0.1$$
$$m(\{John\}) = 0.1$$
$$m(\{Carol\}) = 0.0$$
$$m(\{Mary, John\}) = 0.1$$
$$m(\{Mary, Carol\}) = 0.2$$
$$m(\{John, Carol\}) = 0.0$$
$$m(\{Mary, John, Carol\}) = 0.5$$

Note, the mass assigned to Mary in $m(\{Mary\})$ only reflects the evidence that points directly to her, so does the evidence assigned to $m(\{John\})$ only reflect the evidence directly pointing at John. The evidence that points to Mary or Carol is only assigned to $m(\{Mary, Carol\})$, as is the evidence towards Mary or John solely represented in $m(\{Mary, John\})$. It is intuitively clear that the evidence provided is not enough to identify the villain and there is still the chance any one of them has committed the crime. Hence, there is still a lot that is not known, which is reflected in the probability mass assigned to $(\{Mary, John, Carol\})$.

After the probability mass has been distributed, belief and plausibility, as defined in [29], can be used to establish how much evidence there is against each person. The belief, for example $Bel(\{Mary\})$, states how strongly the evidence supports the theory that Mary has committed the crime. The plausibility $Pl(\{Mary\})$ denotes how much evidence there is that does not contradict the theory that Mary has committed the crime.

$$Bel(\{Mary\}) = m(\{Mary\}) = 0.1$$
$$Pl\{(Mary\}) = m(\{Mary\}) + m(\{Mary, Carol\}) + m(\{Mary, John\}) +$$
$$m(\{Mary, John, Carol\}) = 0.9$$

The analysis of the evidence for all three suspects provides:

$$Bel(\{Mary\}) = 0.1$$
$$Bel(\{John\}) = 0.1$$
$$Bel(\{Carol\}) = 0.0$$

$$Pl(\{Mary\}) = 0.9$$
$$Pl(\{John\}) = 0.7$$
$$Pl(\{Carol\}) = 0.7$$

Together, belief and plausibility provide an interval of probability mass. The real probability for each person to have committed the crime lies somewhere within this interval but, based on the provided evidence, it cannot be determined any further.

$$Mary = [0.1, 0.9]$$
$$John = [0.2, 0.7]$$
$$Carol = [0.0, 0.7]$$

However, if probabilistic values are needed, the *pignistic transformation*, as described in [29], provides a way to calculate a probability distribution based on the mass distribution. Note, that this is only an estimation of the real probability value that is called the *pignistic probability (BetP)* (e.g. [27]). In this example it provides:

$$BetP(Mary) \approx 0.42$$
$$BetP(John) \approx 0.32$$
$$BetP(Carol) \approx 0.26$$

4.5.1   Evidence Combination Within Evidence Theory

Often, evidence stemming from different sources needs to be combined. In the above example of crime investigation, this could be the case when several investigators decide to work together. Combining evidence in the form of mass functions can be done by Dempster's rule of combination [9]:

$$m_{1,2}(A) = \frac{\sum_{\substack{B \cap C = A \\ B,C \subseteq \Omega_X}} m_1(B)m_2(C)}{1 - K} \tag{4.1}$$

where K is a measure of the conflict between the two mass functions and is defined as:

$$K = \sum_{B \cap C = \emptyset} m_1(B)m_2(C) \tag{4.2}$$

Dempster's rule of combination is both, commutative and associative and builds on the assumption that all sources of evidence (the mass functions that are to be combined) are conditionally independent of each other. For the crime investigation example this means that the two investigators must base their respective mass functions on completely different information sources and, for example, cannot use the same witness reports. As a demonstration of the formula consider the two mass functions m_1 and m_2 that relate to:

$$m_1(\{Mary\}) = 0.1$$
$$m_1(\{John\}) = 0.1$$
$$m_1(\{Carol\}) = 0.0$$
$$m_1(\{Mary, John\}) = 0.1$$
$$m_1(\{Mary, Carol\}) = 0.2$$
$$m_1(\{John, Carol\}) = 0.0$$
$$m_1(\{Mary, John, Carol\}) = 0.5$$

$$m_2(\{Mary\}) = 0.2$$
$$m_2(\{John\}) = 0.1$$
$$m_2(\{Carol\}) = 0.0$$
$$m_2(\{Mary, John\}) = 0.2$$
$$m_2(\{Mary, Carol\}) = 0.1$$
$$m_2(\{John, Carol\}) = 0.0$$
$$m_2(\{Mary, John, Carol\}) = 0.4$$

The conflict K between the two is: 0.06 which leads the nominator in Dempster's rule of combination to be $1-K = 0.94$. With Dempster's rule of combination the combined evidence expressed in the mass function $m_{1,2}$ is:

$$m_{1,2}(\{Mary\}) \approx 0.3$$
$$m_{2,1}(\{John\}) \approx 0.1$$
$$m_{1,2}(\{Carol\}) \approx 0.0$$
$$m_{1,2}(\{Mary, John\}) \approx 0.2$$
$$m_{1,2}(\{Mary, Carol\}) \approx 0.2$$
$$m_{1,2}(\{John, Carol\}) \approx 0.0$$
$$m_{1,2}(\{Mary, John, Carol\}) \approx 0.2$$

The pignistic transformation on this results in:

$$Bet\,P(Mary) \approx 0.57$$
$$Bet\,P(John) \approx 0.27$$
$$Bet\,P(Carol) \approx 0.16$$

It can be interesting to compare this with the results from Bayesian fusion, given in the next section. In order to do that the underlying mass and probability functions must be comparable. This can be achieved, when the values for the probability functions match the BetP values established from the mass functions, as is the case in the example.

Dempster's rule strongly emphasizes the agreement between the different sources. However, depending on the task at hand, this might not always be the appropriate way to combine evidence. Hence, there exist some examples in the literature where the operator yields counter intuitive results, e.g. [33]. However, there also are ways to handle the problematic situations as has been shown by [12]. Furthermore, several alternative combination rules have been developed (see for example [23]) that represent and/or combine evidence in a slightly different way than originally presented by Dempster [9] and Shafer [24] and thereby avoid the problem. In the example (e.g. [33]), the frame of discernment consists of three possible illnesses:

$$\Omega_X = \{meningitis, concussion, tumor\}$$

Two medical experts each provide a diagnosis based on some evidence that they collected according to their diagnosis methods. The diagnosis of expert 1 (m_1) strongly indicates that the patient has meningitis, whereas expert 2's diagnosis (m_2) suggest that the patient most likely has a concussion.

$$m_1(\{meningitis\}) = 0.99$$
$$m_1(\{concussion\}) = 0.00$$
$$m_1(\{tumor\}) = 0.01$$

$$m_2(\{meningitis\}) = 0.00$$
$$m_2(\{concussion\}) = 0.99$$
$$m_2(\{tumor\}) = 0.01$$

The combined result with Dempster's rule of combination yields:

$$m_{1,2}(\{meningitis\}) = 0.0$$
$$m_{1,2}(\{concussion\}) = 0.0$$
$$m_{1,2}(\{tumor\}) = 1.0$$

This result appears to be counterintuitive, as common sense would indicate that the patient has either a concussion or meningitis. However, in other scenarios, this way of fusing evidence might be exactly what is needed, for instance when the motto:

Once you eliminate the impossible, whatever remains, no matter how improbable, must be the truth.

by Sir Arthur Conan Doyle is applicable as, for example, in the crime investigation example from above. If two investigators come to the following evidence distributions:

$$m_1(\{Mary\}) = 0.99$$
$$m_1(\{John\}) = 0.00$$
$$m_1(\{Carol\}) = 0.01$$

$$m_2(\{Mary\}) = 0.00$$
$$m_2(\{John\}) = 0.99$$
$$m_2(\{Carol\}) = 0.01$$

where the assignment of 0.00 could represent that the person in question has a definite aliby and the combined result becomes:

$$m_{1,2}(\{Mary\}) = 0.0$$
$$m_{1,2}(\{John\}) = 0.0$$
$$m_{1,2}(\{Carol\}) = 1.0$$

this is what is expected.

Hence, in this application area a result like this, stemming from a so called *conjunctive* combination rule that emphasizes the agreement between the sources and eliminates the alternatives where the two sources are in conflict, is wanted. In another application area the conflict between the two sources might need to be treated in a different way, which can be done using a different fusion rule.

A *cumulative* fusion rule does not exclude any evidence in the combination and can, for example, be useful when information from different sensors need to be integrated, e.g. in order for a robot to keep its balance. An *averaging* fusion rule is applicable when information from homogeneous sources is combined to achieve an overall better accuracy. In the medical diagnosis example above it would be reasonable to apply an averaging fusion rule.

4.5.2 Evidence Combination with Bayesian Theory

Bayesian theory can also be utilized to fuse evidence stemming from different sources [2, 3, 19]. Here, the evidence is presented in the form of likelihoods (as described in [29]). Assuming that investigator 1 has made an observation y, with a corresponding likelihood $p(y|X)$, and investigator 2 has made an observation z with a corresponding likelihood $p(z|X)$, the joint likelihood of the two observations can be formulated by:

$$p(y, z|X) = p(z|X)p(y|X), \tag{4.3}$$

which utilizes the assumption of *conditional independence* between y and z given that the true state of X is known. In order to obtain a probability function as a result, the above equation can be normalized to:

$$\hat{p}(y, z|X) = \frac{p(z|X)p(y|X)}{\displaystyle\sum_{x \in \Omega_X} p(z|x)p(y|x)} \tag{4.4}$$

In terms of the previous example, the frame of discernment is defined as:

$$\Omega_X = \{Mary, John, Carol\}$$

Assuming that the two investigators have made the following observations and constructed normalized likelihoods according to:

$$p(y|Mary) = 0.5$$
$$p(y|John) = 0.3$$
$$p(y|Carol) = 0.2$$

$$p(z|Mary) = 0.4$$
$$p(z|John) = 0.3$$
$$p(z|Carol) = 0.3$$

Taken together these observations intuitively constitute that most evidence is pointing towards Mary since the likelihoods for her are the highest in both observations. Using Eq. (4.4), the following combined result can be obtained:

$$\hat{p}(y, z|Mary) \approx 0.57$$
$$\hat{p}(y, z|John) \approx 0.26$$
$$\hat{p}(y, z|Carol) \approx 0.17$$

This can be seen as a reinforcement of the evidence towards Mary when compared to the two original operand likelihoods $p(y|\cdot)$ and $p(z|\cdot)$. Hence, Bayesian fusion is a conjunctive combination rule that reinforces the agreement of the obtained observations.

4.6 Uncertainty

Uncertainty management plays a vital role within the information fusion process where performance is often measured in terms of accuracy, confidence and timeliness [8]. Uncertainty arises at many different places in the process and for many different reasons and is often distinguished into *epistemic* uncertainty and *aleatory* uncertainty.

Epistemic uncertainty describes uncertainty that originates form a lack in accuracy, e.g. of a measuring process. It is also called subjective uncertainty, reducible uncertainty, state of knowledge uncertainty, ignorance, or type B uncertainty [23]. Epistemic uncertainty is often known and can be articulated, e.g. the measured length of an object might be off by ±5 millimeters. It could, in principle, be reduced by using a more accurate measuring tool. Aleatory uncertainty refers to uncertainty that stems from a random process, hence it is also called randomness as in [29], stochastic uncertainty, type A uncertainty, irreducible uncertainty, variability or objective uncertainty [23]. Rolling a die is one example. Aleatory uncertainty can not be eliminated.

Depending on both, the type of uncertainty and the task to fulfill, uncertainty is often interpreted differently, and the distinction of different types of uncertainty makes it convenient for humans to quickly describe where the uncertainty originated from [32]. However, from a mathematical perspective, all uncertainty stems from a lack of knowledge and can therefore, mathematically, be treated the same [32].

Nevertheless, depending on the application tasks, different ways in that the uncertainty is represented and handled are necessary [26]. Hence, it needs to be carefully chosen which of the uncertainty management methods will be used to model, combine, and reason with uncertainty. Some of the most popular methods are probability theory (Bayesian theory), fuzzy sets, evidence theory (belief functions), non-

Fig. 4.5 Types of uncertainty (adapted from Klir and Yuan [22])

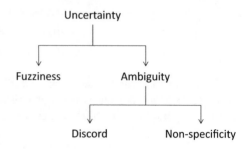

monotonic logic, and possibility theory. The foundations of Bayesian theory, fuzzy set theory and evidence theory are described in another chapter in this book ([29]) and evidence combination within evidence theory and Bayesian theory are described in the previous section. An overview of which type of uncertainty is supported by which type of theory can be found in [26].

There are several taxonomies of uncertainty to be found in the literature and a review of many of them is presented in [18]. In this chapter the taxonomy by Klir and Yuan [22] shown in Fig. 4.5 is adopted. Here, uncertainty is subdivided into two main types: *fuzziness* and *ambiguity*. Uncertainty is referred to as fuzziness when sharp distinctions between different alternatives are lacking. For example, the distinction between a chair and an armchair could be regarded as being fuzzy. Ambiguity describes the situation when there are more than one alternatives, but it is unclear which of the alternatives is the truth. Ambiguity can further be divided into *non-specificity* and *discord* [22]. Non-specificity describes a set of known alternatives, e.g. the outcome of the rolling of the die, but it cannot, beforehand, be said which of those alternatives it will be. Discord, also known as *conflict* or *entropy*, on the other hand, describes the case where the uncertainty stems from conflicting information.

The task of handling uncertainty is twofold. Firstly, despite the incoming information being uncertain, the best and as correct as possible assessment of the situation at hand is what the process should arrive at. Secondly, it is of interest how certain or uncertain this assessment is. In order to assure that the fused result of incoming information is as good as possible, it also might be of interest to know how uncertain the incoming pieces of information are before they are fused. That way, it would be possible to disregard incoming information and exclude them from the fusion process if they are considered to be too uncertain.

Uncertainty is usually expressed by an *uncertainty function* such as the probability measure p in probability theory, the mass function m, the believe function Bel or the plausibility function Pl in evidence theory, which have all been defined in [29]. In order to measure uncertainty, a so called *uncertainty measure* that assigns a non-negative real number to the uncertainty function can be used [21].

The uncertainty within the probability function p can be measured by the *Shannon entropy* [25]:

$$H(p(X)) = - \sum_{x \in \Omega_X} p(x) \log_2(p(x)) \tag{4.5}$$

The function H reaches a maximum of uncertainty at the uniform probability distribution and a minimum of uncertainty when all probability mass resides on a single state.

Within evidence theory, the uncertainty of the mass function m can, for example, be measured by the *Aggregate Uncertainty (AU)*. As it is possible in evidence theory to measure discord and non-specificity separately, the name Aggregated Uncertainty reflects that AU measures both taken together; hence, it is also called a measure for the *total uncertainty (TU)*. There are several measures for total uncertainty available in the literature, e.g. [17, 20, 21], but AU is the only one that fulfills all requirements proposed by [16] to be necessary for an uncertainty measure to fulfill. Formally, AU is defined as in [16]:

$$AU(m) = \max_{p(X) \in \mathscr{P}(X)} H(p(X)), \tag{4.6}$$

where H denotes the Shannon entropy as defined by Eq. (4.5) and where:

$$\mathscr{P}(X) = \left\{ p(X) : Bel(A) \leq p(A) \leq Pl(A), \right.$$
$$\left. \sum_{x \in \Omega_X} p(x) = 1, A \subseteq \Omega_X \right\}, \tag{4.7}$$

is the set of all probability distributions that are consistent with Bel and Pl defined on m.

References

1. URL: http://vestibular.org/understanding-vestibular-disorder/human-balance-system
2. Arnborg, S. (2004). Robust Bayesianism: Imprecise and paradoxical reasoning. In *Proceedings of the 7th International Conference on Information Fusion* (2004)
3. Arnborg, S. (2006). Robust Bayesianism: Relation to evidence theory. *Journal of Advances in Information Fusion, 1*(1), 63–74.
4. Bae, J., Falkman, G., Helldin, T., & Riveiro, M. (2018). Visual data analysis. In A. Said, & V. Torra (Eds.), *Data Science in Practice* (2018)
5. Blasch, E., Bossé, E., & Lambert, D. (2012). *High-level information fusion management and system design* (1st ed.). Norwood, MA, USA: Artech House Inc.
6. Blasch, E., & Plano, S.: DIFG level 5 (user refinement) issues supporting situational assessment reasoning. In *International Conference on Information Fusion* (2005)
7. Boyed, J.: The essence of winning and losing. URL http://dnipogo.org/john-r-boyd/
8. Costa, P. C. G., Laskey, K. B., Blasch, E., & Jousselme, A. L.: Towards unbiased evaluation of uncertainty reasoning: The UREF ontology. In *International Conference on Information Fusion* (2012)
9. Dempster, A. P. (1969). A generalization of Bayesian inference. *Journal of the Royal Statistical Society*, 205–247. Wiley-Blackwell (1969)
10. Duarte, D., & Ståhl, N. (2018) Machine learning: A concise overview. In A. Said, & V. Torra (Eds.),*Data Science in Practice*

11. Endsley, M., & Kiris, E. (1995). The out-of-the-loop performance problem and level of control in automation. *Human Factors: The Journal of the Human Facors and Ergonomics Society*, *37*, 381–394.
12. Haenni, R.: Shedding new light on Zadeh's criticism of Dempster's rule of combination. In *International Conference on Information Fusion*, pp. 879–884
13. Hall, D. L., & Llinas, J. (2001). *Handbook of Multisensor Data Fusion*. CRC Press LLC
14. Hall, M., & McMullen, S. (2004). *Mathematical Thechniques in Multisensor Data Fusion*. Artech House
15. Hand, D. J., Smyth, P., & Mannila, H. (2001). *Principles of Data Mining*. Cambridge, MA, USA: MIT Press.
16. Harmanec, D., & Klir, G. J. (1994). Measuring total uncertainty in Dempster-Shafer theory: A novel approach. *International Journal of General Systems*, 405–419. Taylor & Francis
17. Jousselme, A. L., Liu, C., Grenier, D., & Bossé, É. (2006). Measuring ambiguety in the evidence theory. *IEEE Transactions on Systems, Man, and Cybernetics*, 890–903
18. Jousselme, A. L., Maupin, P., & Bossé, É. (2003). Uncertainty in a situation analysis perspective. In *International Conference of Information Fusion*, pp. 1207–1214
19. Karlsson, A., Johansson, R., & Andler, S. F. (2011). Characterization and empirical evaluation of bayesian and credal combination operators. *Journal of Advances in Information Fusion*, *6*, 150–166.
20. Klir, G. J. (2003). An update on generalized information theory. In *International Symposium on Imprecise Probability: Theories and Applications*
21. Klir, G. J., & Smith, R. M. (2001). On measuring uncertainty and uncertainty-based information: Recent developments. *Annals of Mathematics and Artificial Intelligence*
22. Klir, G. J., & Yuan, B. (1995). *Fuzzy sets and fuzzy logic: theory and applications*. PTR, Upper Saddle River, NJ: Prentice Hall.
23. Sentz, K., & Ferson, S. (2002). *Combination of evidence in Dempster-Shafer theory*. SANDIA: Tech. rep.
24. Shafer, G. (1976). *A mathematical theory of evidence*. Princeton University Press
25. Shannon, C. E. (1948). A mathematical theory of communication. *The Bell System Technical Journal 27*, 379–423, 623–656 (1948)
26. Smets, P. (1999). Imperfect information: Imprecision—Uncertainty. In *UMIS*
27. Smets, P. (2000). Data fusion in the transferable belief model. In *Third International Conference on Information Fusion*
28. Steinberg, A., Bowman, D., & White, F. (1999). Revisions to the JDL data fusion model. In *Sensor Fusion: Architechtures, Algorithms and Applications*
29. Torra, V., Karlsson, A., Steinhauer, H. J., & Berglund, S. (2018). Artificial intelligence. In A. Said, & V. Torra (Eds.), *Data Science in Practice*
30. Waltz, E. L. (1998). Information understanding: integrating data fusion and data mining processes. In *Circuits and Systems*. ISCAS'98. Proceedings of the 1998 IEEE International Symposium on, vol. 6, pp. 553–556. IEEE (1998)
31. Waltz, E. L., & Llinas, J. (1990). *Multisensor data fusion*. Norwood, MA, USA: Artech House Inc.
32. Winkler, R. L. (1996). Uncertainty in probabilistic risk assessment. In *Reliability Engineering and System safety*, pp. 127–132
33. Zadeh, L. A. (1984). Review of books: A mathematical theory of evidence. *AI Magazine, 5*, 81–83.

Chapter 5
Information Retrieval and Recommender Systems

Alejandro Bellogín and Alan Said

Abstract This chapter provides a brief introduction to two of the most common applications of data science methods in e-commerce: information retrieval and recommender systems. First, a brief overview of the systems is presented followed by details on some of the most commonly applied models used for these systems and how these systems are evaluated. The chapter ends with an overview of some of the application areas in which information retrieval and recommender systems are typically developed.

5.1 Introduction

Information retrieval is the process of retrieving information relevant to a query from an information source, e.g. a book from a library based on a title, or a relevant search result based on a query posted to a web search engine. Recommender systems, closely related to information retrieval systems, however work *without* a query. Instead, the recommender system attempts to identify the most relevant piece of information solely based on an implicitly expressed information need and intent—i.e., the user profile.

In the Information Retrieval (IR) community, information retrieval often means *text retrieval* by default, either intentionally or unintentionally. This might be due to historical reasons [40] or simply because text retrieval has been the most predominant information retrieval application. But, nonetheless, there exist many other forms of information retrieval applications. A typical example is collaborative filtering, which aims at finding information items a target user is likely to like by taking into account other users' preferences or tastes. Unlike text retrieval, collaborative filtering (CF) does not necessarily need textual descriptions of information items and user needs.

A. Bellogín
Iniversidad Autñoma de Madrid, Madrid, Spain
e-mail: alejandro.bellogin@uam.es

A. Said (✉)
University of Skövde, Skövde, Sweden
e-mail: alansaid@acm.org

© Springer International Publishing AG, part of Springer Nature 2019
A. Said and V. Torra (eds.), *Data Science in Practice*, Studies in Big Data 46,
https://doi.org/10.1007/978-3-319-97556-6_5

It makes personalized recommendations by aggregating the opinions and preferences of previous users. Originally, the idea of collaborative filtering was derived from heuristics and coined by Goldberg and colleagues when developing an automatic filtering system for electronic mail [17]. Due to the uniqueness of the problem, it has been modeled and studied differently since then, mainly drawing from the preference prediction and machine learning view point [6].

In this chapter, we introduce the main concepts and models from Information Retrieval and Recommender Systems, paying special attention to one problem where there are several connections between these two areas: evaluation. We also present some of the most important applications for each community.

5.2 Models

5.2.1 Information Retrieval Models

Text retrieval techniques have been widely used in many different areas such as web retrieval [2], image and video retrieval [41], or content-based recommendation [42]. Among many techniques, one of the most commonly used are the term weighting functions, from which several extensions and probabilistic adaptations have been proposed throughout the years; the most representative methods are reviewed in the next sections.

5.2.1.1 Classical Retrieval Models: Boolean, TFIDF, Vector Models

The Boolean retrieval model was used in the first search engines and is still in use today [13]. Documents are retrieved if they exactly match the query terms, however, this does not generate a ranking, since this model assumes that all documents in the retrieved set are equivalent in terms of relevance. Nonetheless, this simple model has some advantages: it is very predictable and easy to explain to users; from an implementation point of view, it is also very efficient because documents can be directly removed from the retrieved set if they do not include any query features.

The major drawback of this approach, however, is the lack of a proper ranking algorithm. Because of this, the vector space model (VSM) was proposed [38], where documents and queries are assumed to be part of the term space. In such space, term weighting functions and similarities between documents can be defined. Usually, the terms are weighted according to their frequency in the document (TF, or term frequency) or normalized with respect to the number of documents where they appear (IDF, or inverse document frequency). Cosine similarity is typically used to compute the similarity between the document and query vectors, generated based on a specific term weighting function. This is used because, even though there is no explicit definition of relevance in the VSM, there is an implicit assumption that relevant documents are located closer in the term space to the query—i.e., the distance between their corresponding vectors should be small.

5.2.1.2   Probabilistic Models: Probabilistic Model, BM25, Language Models

Probabilistic models, in contrast with the previously described retrieval models, provide some levels of theoretical guarantees that the models will perform well, as long as the model assumptions are consistent with the observed data. The Probability Ranking Principle or PRP [35] states that ranking the documents by probability of relevance will maximize the precision at any given rank—assuming that the relevance of a document to a query is independent of other documents. Since the PRP does not tell us how to calculate or estimate such probability of relevance, each different probabilistic model proposes a different method for that which, at the end, could produce better or worse documents for a given query.

One of the first methods to calculate the probability of relevance using document terms was proposed in [36]. Other approaches estimate text statistics in the documents and queries and then build up the term weighting function through them [3]. The BM25 ranking algorithm extends the scoring function to include document and query term weights [34]:

$$\sum_{t_i \in Q} \log \frac{(r_i + 0.5)/(R - r_i + 0.5)}{(n_i - r_i + 0.5)/(N - n_i - R + r_i + 0.5)} \cdot \frac{(k_1 + 1)f_i}{K + f_i} \cdot \frac{(k_2 + 1)qf_i}{k_2 + qf_i}$$

(5.1)

where f_i is the frequency of term t_i in the document, qf_i is the frequency of the term t_i in the query, n_i (r_i) is the number of (relevant) documents that contain term t_i, N (R) is the number of (relevant) documents in the collection, and k_1, k_2, K are free parameters. While this extension is based on probabilistic arguments and experimental validation, it is not a formal model. Nonetheless, it performs very well in many different tasks.

More sophisticated approaches like the Relevance-Based Language Models (or Relevance Models for short, RM) are among the best-performing ranking techniques in text retrieval [22]. They were devised with the aim of explicitly introducing the concept of relevance, intrinsic to the probabilistic model, in statistical Language Models (LM) [30]. In common IR settings, the exact and complete set of relevant documents is generally unknown; relevance feedback techniques work with approximations to this set, which can be obtained by a wide variety of approaches, such as asking the user (explicit relevance feedback), or just taking the initial output of a well-performing IR system as a good guess (pseudo relevance feedback). Given a query and such an approximation to the set of relevant documents, RM selects good expansion terms from those present in the pseudo-relevant documents in order to formulate and run a better query.

5.2.1.3 Models for the Web: PageRank and Learning to Rank

Ranking documents in the Web is substantially different from ranking a static collection, since other aspects besides the relevance to a query should be considered: spam should be avoided, pages with higher quality should be presented before (authorities), new content is created every day (hence, the models should be efficient so they are executed fast and able to manage huge amounts of data), and so on [5].

One of the most important signals available in the Web is the links between Web pages. The link structure of the Web allows to classify the pages into authorities, hubs, sinks, and normal pages by simply checking whether they are linked by many pages (authority), link to important pages (hubs), do not have further links (sinks), and the rest. PageRank and HITS algorithms exploit these ideas and compute a score for every page in the Web graph, by simulating a random surfer that navigates the Web [7, 20].

Another important signal that is typically exploited by search engines is the information about which documents are being clicked for each query. Once enough information is collected (click logs), standard machine learning algorithms can be used to learn a ranking based on some training data. The loss function to minimize is either the number of mistakes done by the algorithm or the (negative) average precision, see [12, 24] for a more comprehensive overview of these techniques.

5.2.2 Recommender System Models

Recommender Systems (RS) are ubiquitously used on the World Wide Web and are increasingly being put to use in more application scenarios. Even though recommender systems have grown to prominence in the *online* world, the techniques for many of today's systems were developed prior to the Internet breakthrough. Traditionally, recommender systems have been closely linked with information retrieval systems as they use similar models to identify relevant data. However, in recent years, with the rapid developments of recommender systems, the two fields have come to grow more separate.

In general, recommender systems approaches are categorized as *collaborative filtering*-based, *content*-based, or as *hybrid* models which blend both collaborative and content information into one unified approach.

5.2.2.1 Collaborative Filtering Recommender Systems

Collaborative Filtering (CF) was initially developed as a means of filtering out relevant news posts on Usenet [32]. The intuition behind the initial *user-based* CF algorithm is that people with similar preferences to each other have not all seen the same items. Thus, by looking at the items that users similar to oneself have seen, we can find items which should be of interest to us.

The traditional user-based collaborative filtering method uses the k-nearest neighbor algorithm in order to find the most suitable recommendations by predicting rating scores that will be given to items by users, i.e.

$$\tilde{r}_{ui} \propto \sum_{v \in N_k(u)} sim(u, v) r(v, i) \tag{5.2}$$

where r_{ui} refers to the predicted rating that user u will assign to item i, $N_k(u)$ is the set of the k most similar users (nearest neighbors) to user u, and $sim(u, v)$ the similarity between user u and neighbor v. The similarity method $sim(u, v)$ can be chosen freely. Common methods are *cosine similarity* or *Pearson correlation*.

Cosine similarity measures the cosine angle between two users where each user is represented by a vector of their ratings, i.e.

$$cos(u, v) = \frac{u \cdot v}{||u|| ||v||} = \frac{\sum_{i=1}^{|I|} u_i v_i}{\sqrt{\sum_i u_i^2} \sqrt{\sum_i v_i^2}} \tag{5.3}$$

where u and v are the two user respectively, and u_i and v_i the ratings to item i. Similarly, the Pearson correlation measures the linear correlation between two vectors, where each vector represents the ratings of a user, i.e.

$$P(u, v) = \frac{\sum_{i \in I_{uv}} (r_{ui} - \bar{r}_u)(r_{vi} - \bar{r}_v)}{\sqrt{\sum_{i \in I_{uv}} (r_{ui} - \bar{r}_u)^2} \sqrt{\sum_{i \in I_{uv}} (r_{vi} - \bar{r}_v)^2}} \tag{5.4}$$

where u and v are two, I_{uv} the intersection of user u's and v's rated items, and r_{ui} and r_{vi} the rating given to item i by user u and user v respectively.

Analog to user-based collaborative filtering, *item-based* collaborative filtering instead looks at similarities between items. Instead of identifying similarities between users, this method attempts to find the most similar items to those the candidate user has already rated (or otherwise interacted with).

User-based and item-based collaborative filtering belong to the *memory-based* family of recommendation approaches. As the name implies, memory-based approaches keep the entire database of user-item interactions in memory in order to find the most appropriate recommendations. *Model-based* recommendation approaches instead create a recommendation model from the available data, when providing recommendations to users, recommendations can be accessed from the pre-computed model. There is a wide variety of memory-based recommendation models, some of the most commonly ones fall into the category known as *matrix factorization*, e.g. Bayesian Personalized Ranking [31], SVD++ [21]. Somewhat related, *probabilistic topic modeling* approaches like Latent Dirichlet Annotation [4] also fall in the model-based category.

5.2.2.2 Content-Based Recommender Systems

Content-based recommender systems (CB) analyze the items the user liked in the past and recommend items having similar features. The main process of making recommendations using a content-based approach consists in matching users preferences and interests obtained in the users' record, with the attributes of the items [25]. Since CB systems recommend items analyzing the user profile, they aim to maximize the similarity between the user profile and the content of the item, that is, $sim(UserProfile(u), Content(i))$, where $Content(i)$ will be the item profile (attributes that characterize item i).

As CB recommenders tend to use articles represented by text, the content is normally represented by keywords using simple retrieval models such as the ones presented before, the most representative example being the Vector Space Model, where cosine similarity with weights computed using either TF or TFIDF are the most popular techniques.

Nonetheless, there are other techniques for CB recommendation. One of the most important techniques is Bayesian classifiers. These approaches estimate the posterior probability $P(c \mid d)$ of a document belonging to a specific class c based on the prior probability of the class $P(c)$, the probability of observing the document in class c (i.e., $P(d \mid c)$) and the probability of observing the document d denoted as $P(d)$ [16]. The estimation of $P(d \mid c)$ is complicated, thus it is common to use the naïve Bayes classifier, as shown in [27] for book recommendation and in [28] for classifying unrated web pages. With the naïve Bayes classifier, the document is replaced by a vector of keywords over the system vocabulary. Each component of the vector indicates whether that keyword appeared in the document or not. If we work with binary values, we are using a multivariate Bernoulli approach and if we count how many times the word appeared in the document, we are making use of multinomial naïve Bayes [16].

5.2.2.3 Hybrid Recommender Systems

Hybrid recommendation takes advantage of the techniques from two or more recommender systems to achieve a higher performance while limiting the potential drawbacks that each system may obtain separately. Although there are hybrids that combine implementations of recommendations of the same type (for example, two content-based techniques), the most interesting ones are those that are able to work with different recommenders. There are different strategies for hybrid recommendation which [9] summarized in the seven methods shown in Table 5.1.

Table 5.1 Hybrid recommendation methods from [9]

Hybrid technique	Description
Weighted	Each recommender obtains a score for each candidate item and these scores are combined using a linear formula
Switching	It switches between recommenders depending on the situation
Mixed	Each recommender makes its own recommendations and the final output is a combination of them
Feature combination	Features derived from various sources are combined and sent to the recommendation scheme
Feature augmentation	Similar to feature combination but instead of deriving, the recommenders augment (compute) new features and send them to the final recommendation scheme
Cascade	They normally use a weak and a strong recommender. The weak recommender is only used when breaking ties in the ranking
Meta-level	A recommender produces a model, which is the input for the second recommender. Similar to feature augmentation but the second recommender does not work with raw data, only with the model provided by the first recommender

5.3 Evaluation

The evaluation of IR and RS is seemingly similar, but has stark conceptual differences. IR evaluation measures how well a system is able to retrieve relevant results to a certain query, whereas evaluation of RS looks into how relevant each recommended item is to the user it gets recommended to. Evaluation of recommender systems has, to a large extent, been based on information retrieval concepts such as precision, recall, F-measure, etc. These measures represent some form of quality of the system, e.g. the higher the precision and/or recall value is, the more accurate the system is. However, even though information retrieval systems and recommender systems are similar both in their use and implementation, there is a distinct contextual difference; whereas retrieving a known (but sought for) item is positive, recommending a known item has far lower utility [19].

There exist a few key concepts in evaluation of both information retrieval and recommender systems that are as relevant to both. One of these, and very likely the most important aspect of both information retrieval and recommendation results, is the *relevance* of the information presented to the end user. Any piece of information retrieved or recommended to the user should exhibit a certain relevance to the users information need. In general, evaluation of both type of systems focuses on measuring the relevance (through some metric) of the items presented to the user.

In order to measure the relevance of items, we need to know the *ground truth*, i.e. we need to know which items are relevant to which users. It is using this ground truth that we can gauge how well a recommender or retrieval system performs. The ground truth is specified by a dataset, for recommendation purposes, this often consists of historical interactions between items and users, for information retrieval it is often found in a *document collection* which consists of e.g. a set of documents, search queries used on the set of documents, and relevance assessments which tell how relevant a document is to each query.

5.3.1 Metrics

Recommendation qualities are commonly expressed through a number of metrics and methods. The choice of these is often based on the type of dataset used in the system, the use case, expected outcome, etc. Arguably the most common metrics in both recommender systems and information retrieval is the precision (P) and recall (R) pair [19]. These metrics are usually applied in offline training/test scenarios, where algorithms are trained using a portion of the available data and then evaluated by comparing predictions to a withheld portion of the data, i.e. *true positive* recommendation.

Precision is the fraction of relevant retrieved documents. In a recommender system evaluation setting it corresponds to the true positive fraction of recommended items. Recall is the fraction of all relevant items which are retrieved. The formula for calculating precision is shown in Eq. 5.5 while recall is shown in Eq. 5.6. In both equations, *relevant* refers to the complete set of relevant items, and *retrieved* refers to the complete set of retrieved items.

$$P = \frac{|\{relevant\} \cap \{retrieved\}|}{|\{retrieved\}|} \tag{5.5}$$

$$R = \frac{|\{relevant\} \cap \{retrieved\}|}{|\{relevant\}|} \tag{5.6}$$

Commonly, precision is expressed as precision at k where k is the length of the list of recommended items, e.g. $P@1 = 1$ would indicate that one item was recommended, and the item was deemed to be a true positive recommendation, $P@2 = 0.5$ would indicate that two items were recommended and one them was deemed a true positive, etc.

Variants of precision used for recommender evaluation include *Average Precision* (AP) and *Mean Average Precision* (MAP). Both these metrics are used when more than one item is recommended. They extend the precision metric by taking into consideration the position of true positive recommendations in a list of recommended items, i.e. the position k in a list of n recommended items in Eq. 5.7. $rel(k; q)$ is a binary classifier taking the value 1 if the item at position k is relevant for query q and 0 otherwise. Mean Average Precision additionally averages the scores at each query (or user, in recommendation), as shown in Eq. 5.8.

$$AP(q) = \frac{\sum_{k=1}^{n} (P@k(q) \times rel(k; q))}{|\{relevant\}|} \tag{5.7}$$

$$MAP = \frac{\sum_{q=1}^{Q} AP(q)}{Q} \tag{5.8}$$

Other common metrics, used in the context of rating prediction, are the Root-Mean-Square Error (RMSE) and normalized Discounted Cumulative Gain (nDCG). In contrast to precision-based metrics, RMSE attempts to estimate the recommendation algorithm's rating prediction error, e.g. by comparing predicted ratings to actual ratings. The lower the error, the better the algorithm performs. RMSE is calculated as shown in Eq. 5.9, where X, Y are two rating vectors (e.g. predicted item ratings vs. actual item ratings), where each position in the vector corresponds to the rating of a specific movie and n the size of the intersection of nonzero elements in both vectors.

$$RMSE(X, Y) = \sqrt{\frac{\sum_{i=1}^{n} (x_i - y_i)^2}{n}} \tag{5.9}$$

nDCG and DCG on the other hand measures the usefulness (relevance) of a document based on its position in the list of recommended items. In a rating scenario, this corresponds to how high the predicted ratings of the top-k items are, the formula is shown in Eq. 5.10 where the *gain* (the predicted rating) of each item i for each user u in a list of J items is represented by g_{ui_j}. nDCG is the DCG over the true DCG, i.e. *ideal* DCG (IDCG)—the actual ratings, as shown in Eq. 5.11.

$$DCG = \frac{1}{N} \sum_{u=1}^{N} \sum_{j=1}^{J} \frac{g_{ui_j}}{\max(1, \log_2 j)} \tag{5.10}$$

$$nDCG = \frac{DCG}{IDCG} \tag{5.11}$$

5.3.2 Collections

As briefly noted in the beginning of this section, document and data collections are a necessity in order to be able to evaluate information retrieval and recommender systems. Even though there are similarities between the collections used for information retrieval and for recommender systems, there are clear differences. Below, we present document collections and historical interaction datasets used for search and recommendation respectively.

5.3.2.1 Document Collections

To ensure fair experimental comparisons and repeatable experimental settings, the data used in such experiments should be fixed. In information retrieval, test collections are assembled consisting of documents, queries, and relevance judgements (for some query-document pairs). Differently to other areas such as Machine Learning, in IR, the queries and the relevance judgments are gathered specifically for a particular search task, at the same time as the documents [13].

In this context, creating relevance judgments require a considerable investment of manual effort. When collections were very small, most of the collections could be evaluated for relevance, hence, the relevance judgments for each query were relatively exhaustive. In larger collections, a technique called *pooling* is used, where the top *k* results from different rankings are merged into a pool, duplicates are removed, and these documents are then presented to the assessors or judges to produce the relevance judgments. This technique is based on two assumptions: most of the documents in the pool are relevant (because they were retrieved high in the ranking by different methods) and those documents not in the pool can be considered not relevant. Both assumptions have been verified to be accurate for some specific IR tasks [5], however, whenever a new retrieval algorithm (that did not contribute to the pool) is evaluated in this way, the effectiveness of this method could be underestimated if it mostly retrieves documents not in the original pool.

Following this evaluation paradigm, information retrieval researchers have built many test collections, ranging from thousands to millions of documents. The Text Retrieval Conferences (TREC) have been the main events around which IR experiments have been designed, usually involving the creation of queries and relevance judgments under a specific track or task (ad hoc, filtering, high precision, diversity, etc.). Other important document collections have been generated in the context of INEX (Initiative for the Evaluation of XML Retrieval), the NTCIR project, and CLEF (Workshop on Cross-Language IR and Evaluation), oriented respectively to XML retrieval, Japanese and cross-lingual retrieval, and different multilingual tasks.

5.3.2.2 Interaction Datasets

When evaluating recommender systems, traditionally, the underlying data that is used contains user-item interactions or relations, these relations can be e.g. a rating given to an item by a user, a record of a user purchasing a product, a record of a user clicking on an item in a list of recommended items, the number of times a user has played a specific song, or how many minutes of a video a specific user played, etc. This type of data is commonly referred to as a user-item matrix U where the rows and columns correspond to users and items respectively. Each cell in the matrix holds the interaction information between the user and the item, i.e. in a movie recommendation scenario each cell will contain either the rating the user has given to the movie or be empty if the user has not rated the movie, in an e-commerce

Table 5.2 A basic example of a user-item matrix where each cell in the matrix corresponds to the rating given by the users to items

	u_1	u_2	u_3	u_4	u_5
i_1	1	3			5
i_2	1			3	5
i_3	2	2			4
i_4	3	4			1
i_5		4	1	5	
i_6		3		5	

interaction matrix each cell will contain e.g. whether or not the user has purchased the item.

Table 5.2 shows an example of a user-item matrix where each cell contains either a rating given to an item by a user, or is empty in the cases where the user has no interaction history with the item—these cases represent the target user-item pairs to be predicted by our recommender system. Note an important difference with respect to how the document collections are built as described before: whereas the relevance of a document with respect to a query can be assessed by any judge (since relevance is usually assumed to be an inherent property of the document, and, hence, it is to some extent universal), in recommendation this assessment is personal, and cannot be inferred by a different user. This leads to a much sparse scenario, where typical values for the density of this matrix (ratio between the number of known cells with respect to the total number of cells) is below 6% [18].

5.4 Applications

Information retrieval and recommender systems have many application domains and are frequently becoming increasingly more intertwined. Below follows an overview of some of the traditional and common application areas of both types of systems.

5.4.1 Information Retrieval

In the Information Retrieval community, information retrieval often means text retrieval by default, either intentionally or unintentionally. This might be due to historical reasons [40] or simply because text retrieval has been the most predominant information retrieval application. But, nonetheless, there exist many other forms of information retrieval applications. In this section we briefly review some of the most important or well-known applications that fall under the umbrella of Information Retrieval [5].

Web search is today the most important application of IR and its techniques [5]. According to [5], there are at least five issues that have impacted how IR methods, technologies, and processes have adapted to the Web: (a) characteristics of the document collection (large, unstructured, connected through hyperlinks, distributed, it has to be collected or *crawled*), (b) size and volume of submitted queries, (c) very large document collection that makes relevance prediction much harder (also because there are many noisy documents), (d) new search tasks and user needs have emerged, together with structured data in both queries and documents, (e) the increasingly pervasive presence of spam on the Web due to new incentives such as advertising and electronic commerce content.

Relevance feedback is the most popular query reformulation strategy. In a relevance feedback cycle, the user is presented with a list of the retrieved documents which are later marked as relevant or not relevant. The main idea consists of selecting important terms from the selected documents and enhance or decrease the importance of these terms in a new query formulation, together with expanding the query with new terms. When using a vector space model, this technique is simply defined as a term reweighting strategy, where different formulations exist to calculate the modified query depending on assumptions on the user behavior regarding the feedback cycle [5, 37].

Before the Web, the design of search tools was targeted to help people write good queries, implying the query language being adopted was usually complex. However, nowadays search engines are used not only to find information but to achieve other goals [5]. The first categorization of queries was done in [8] into three classes: informational, navigational (finding a Web site for browsing), and transactional (interactive tasks such as buying goods or downloading a file). This taxonomy was later refined; moreover, now the focus is on automatically predicting the **query intent**, so that different query attributes are analyzed and selected to derive which of them may be linked to each intent. However, ambiguous queries present a scenario where the user intent is harder to be predicted correctly, as expected. In this context, several proposals have been studied to measure the query difficulty, closely related to its ambiguity. More specifically, the *clarity score* measures how closely related are the documents returned for a query with respect to a particular document collection, which aims to measure the ambiguity of a query towards a collection [11].

Text classification corresponds to a broad problem—mostly addressed by Machine Learning researchers—where a collection of documents is assigned one (or more) class/label out of a predefined number of classes/labels. A particular, and one of the most important, variant of this problem is the topic classification task, where each class describes a topic referred to in the documents. There have been proposed several algorithms to address this problem, either for the multi-label or the single-label scenario—the latter is acknowledged to be harder, because it is also necessary to decide which class is the best one to a given document. Regarding the algorithms, both supervised (decision trees, nearest neighbors, naive Bayes, SVMs) and unsupervised (clustering, direct match) techniques can be applied, although in general supervised algorithms achieve better results, at the expense of requiring available training data.

Text compression allows to represent the text in fewer bits or bytes. Compression methods create a reduced representation, that can be later used to reconstruct the original text, by identifying and using patterns that exist in the text. These techniques help reducing costs associated with space requirements, input/output overhead, and communication delays, since compressed text requires less storage space, takes less time to be transmitted over a communication link, and takes less time to search directly the compressed text. All these advantages come at the expense of more time needed to code and decode the text. One of the most important challenges that appear when text compression techniques are applied in IR systems is the necessity of this type of systems to access text randomly, since most compression methods need to decode the entire text to find a match.

Rank fusion or rank aggregation is needed when you want to combine various result lists from different sources into one list, with no knowledge neither of the process followed by each source to produce those lists nor the data that has been used or the score rank for each element. Examples where rank fusion takes place include, for instance, metasearch, personalized retrieval (combine personalized results with query-based results), multi-criteria retrieval, etc. [39].

Enterprise search refers to the application of information retrieval technologies to information finding within organizations; in particular, of digital documents owned by the organization, such as their external Web site, the company intranet, and any other electronic text (email, database records, reports, shared documents, etc.). According to [5], a far-from-complete list of search-supported tasks that can be found in an enterprise is the following: approving an employee travel request, responding to calls in a call center, responding in the course of a dispute, writing a proposal, obtaining and defending patents, selling to an existing customer, expertise finding, and operating an E-commerce site.

Indexing is the process of creating the data structures needed to enable fast searching. Index creation must be efficient, both in terms of time and space, where usually a tradeoff must be met. Furthermore, indexes must also be able to be efficiently updated when new documents are found. Inverted files are the most common—and best choice for most applications—form of index used by search engines. They contain a list for every index term of the documents that contain such index term. Other techniques not so popular nowadays are suffix arrays and signature files.

Web crawling allows to automatically download Web pages, by means of programs usually called web *crawlers*, *spiders*, or *bots*. The order in which the URLs are traversed in the Web (by following the hyperlinks available in the pages) is important. In general, it is advised to do it using a breadth first strategy, since this is linked to finding *good* sites (i.e., those with higher Pagerank values) sooner in the process; however, depending on the final application and goal of the crawling, other preferences might be considered [23]. One of the main challenges on this topic is how to process efficiently as many pages as possible while preserving resource policies (collectively referred as crawler etiquette, e.g., comply with the Robot Exclusion Protocol) and ethical considerations.

XML (structured) retrieval supports querying and manipulating XML data by using languages that describe the hierarchical structure of XML data instead of simpler models, enough to work with relational databases and unstructured documents [13]. Hence, the focus of these methods is on exploiting the document structure, thanks to the definition of a complex query language, XQuery. It is worth mentioning that the INEX[1] project has allowed to study the extent to which structure is useful in queries, by defining several tasks and building test collections for further evaluation, in a similar way as TREC did for Web search.

Multimedia retrieval is widely recognized as one of the most promising fields in the area of information management [5]. The most important characteristic of this type of systems is the variety of data it must be able to support, such as text, images (still and moving), graphs, and sound. For this reason, the corresponding data model, query language, and access and storage mechanisms should support objects with a very complex structure.

5.4.2 Recommender Systems

In the recommender systems community, recommendation was traditionally separated into two distinct use-cases, top-N recommendation and rating prediction. Top-N recommendation, as the name suggests focuses on generating a list of the N most relevant items to recommend to a specific user, whereas rating prediction focuses on predicting the rating a user will give to a certain item. The rating prediction task was very extensively researched in the past, this was in part due to the Netflix Prize[2] which focused on rating prediction and took place between 2006 and 2009. Today, various top-N aspects of recommendation are at the focus of the recommender systems research and practitioner communities.

Movie recommendation was the primary application of RS systems previously. In part due to the Netflix Prize mentioned above, but also due to the availability of open research datasets such as Movielens [18] and, more recently, MovieTweetings [14]. Traditionally, movie recommendation focused on rating prediction but has gradually shifted to a top-N recommendation use case based on implicit data and models. The shift occurred as movie recommendation become increasingly more applied in streamed video use cases.

Music recommendation is a common application scenario for recommendation. Similar to multimedia retrieval, a music recommender system must be able to support not only collaborative data (the user-item interactions) but also information such as timbre, tempo, genre, etc. making the data and recommendation models more complex. Music recommendation traditionally focuses on one of two tasks: recommendation of a single song or artist, where the recommendation is seen as a standalone action; and playlist recommendation where recommendations should meet requirements posed by the playlist that is being extended and the user who is listening.

---

[1] Initiative for the Evaluation of XML Retrieval, http://inex.mmci.uni-saarland.de.
[2] http://www.netflixprize.com.

User recommendation has become very common with the rise of social media. Recommending acquaintances, colleagues, potential romantic partners, and similar applications differ from traditional recommendation of items in one very significant respect, namely reciprocity. When recommending a social connection to a user, the recommendation should be of value for both of the involved parties [29]. This requirement, although not unique for user recommendation, is a key factor of importance in this setting.

Context-aware recommendation attempts to tailor recommendations for the user's current situation, or context, taking into consideration aspects such as the relevance variability of certain items in certain situations. Consider, for example, an e-commerce recommender system recommending clothing. When recommending clothing in summer, the recommendation could take into consideration the outdoor temperate, and similarly so in winter. The methods for using context in recommendation are generally considered to be either based on contextual filtering or contextual modeling [1], filtering referring to the concept of applying a filter to the list of recommended items to only show the items valid in the specific context, whereas modeling refers to when the recommendation algorithm takes the context into consideration when identifying items to recommend.

Explanations of recommendations tell the user why a certain item is recommended. Recommendation explanation can be generated in various ways, commonly, the explanations are high-level enough to be understood by users who have no insight into how recommendation algorithms work. A simple example could be, e.g. telling the user that an item is recommended because "people similar to you liked this item". Similarity in this case could be based on any number of factors, commonly this can refer to e.g. the collaborative similarity presented in Sect. 5.2.2 of this chapter.

Recommender systems in education is the application of recommendations in order to enhance learning by supporting students through tailored educational materials and exercises according to their learning preferences, knowledge levels, goals, etc. [26]. Recommendation models in this application space must adhere to learning-oriented requirements specified by teachers and educators, e.g. learning materials, expected attained knowledge, and course curricula among others.

Recommendations for groups specifically attempt to tailor recommendation for a group of people at the same time, instead of the regular approach where personalized recommendations are delivered to a single user. Approaches for group recommendations take into consideration issues regarding the group satisfaction with a certain recommendation, e.g. whether it is, say, better to recommend very suitable items for the majority of the group at the cost of leaving certain group members unhappy (maximize happiness), or whether it is better to recommend items that the whole group will be comfortable with, albeit not specifically happy (minimize discontent) [15].

Cross-domain recommendations capture the preferences users express about items in one domain, e.g. movies, and use these preferences to identify relevant recommendations in a separate domain, e.g. music. Given that items, interactions, and preferences can be expressed differently across different domains, cross-domain recommender systems attempt to leverage and transfer the knowledge in other domains and apply on the current one [10]. A well-known example of cross-domain recommendation is e-commerce, e.g. recommendations at Amazon or eBay are often sets of items from various domains, e.g. books, electronics, clothing, etc.

5.5 Summary

This chapter has presented a brief overview of Information Retrieval and Recommender Systems, two very common, if not the most common, applications of data science methods. While information retrieval and recommender systems share many similarities, there are certain differences that need to be addressed when planning, implementing, and deploying them.

Information retrieval and recommender systems are ubiquitously present on the World Wide Web supporting users to find, retrieve, and suggest information. Services ranging from search engines (Google), multimedia delivery (Netflix, Spotify), e-commerce (eBay, Amazon), and social networks (LinkedIn, Facebook, Twitter) have business models closely intertwined with how well their search and recommendation systems work. This very direct connection to commercial real-world applications makes research and development of information retrieval and recommender systems very tightly connected with industry.

References

1. Adomavicius, G., & Tuzhilin, A. (2015). Context-aware recommender systems. In *Recommender systems handbook* (pp. 191–226). Springer.
2. Agichtein, E., Brill, E., & Dumais, S. (2006) Improving web search ranking by incorporating user behavior information. In *Proceedings of the 29th annual international ACM SIGIR conference on Research and development in information retrieval*, SIGIR 2006 (pp. 19–26). New York, NY, USA: ACM. https://doi.org/10.1145/1148170.1148177.
3. Amati, G., & Van Rijsbergen, C. J. (2002). Probabilistic models of information retrieval based on measuring the divergence from randomness. *ACM Transactions on Information Systems* 20(4), 357–389. https://doi.org/10.1145/582415.582416.
4. Amatriain, X., & Pujol, J. M. Data mining methods for recommender systems. In Ricci et al. [33] (pp. 227–262). https://doi.org/10.1007/978-1-4899-76376_7.
5. Baeza-Yates, R. A., & Ribeiro-Neto, B. A. (2011). *Modern information retrieval-the concepts and technology behind search* (2nd ed.). Harlow, England: Pearson Education Ltd. http://www.mir2ed.org/.
6. Breese, J. S., Heckerman, D., & Kadie, C. (1998). Empirical analysis of predictive algorithms for collaborative filtering. In *Proceedings of the fourteenth conference on uncertainty in artificial intelligence*, UAI 1998 (pp. 43–52). San Francisco, CA, USA: Morgan Kaufmann Publishers Inc. http://dl.acm.org/citation.cfm?id=2074094.2074100.
7. Brin, S., & Page, L. (1998). The anatomy of a large-scale hypertextual web search engine. *Computer Networks* 30(1–7), 107–117. https://doi.org/10.1016/S0169-7552(98)00110-X.
8. Broder, A. Z. (2002). A taxonomy of web search. *SIGIR Forum 36* (2), 3–10. http://doi.acm.org/10.1145/792550.792552.

9. Burke, R. (2002). Hybrid recommender systems: Survey and experiments. *User Modeling and User-Adapted Interaction 12*(4), 331–370.
10. Cantador, I., Fernández-Tobías, I., Berkovsky, S., Cremonesi, P. (2015). Cross-domain recommender systems. In *Recommender systems handbook* (pp. 919–959). Springer.
11. Carmel, D., & Yom-Tov, E. (2010). Estimating the query difficulty for information retrieval. Synthesis Lectures on Information Concepts, Retrieval, and Services. Morgan & Claypool Publishers. https://doi.org/10.2200/S00235ED1V01Y201004ICR015.
12. Chakrabarti, S. (2007). Learning to rank in vector spaces and social networks. *Internet Mathematics 4*(2), 267–298. https://doi.org/10.1080/15427951.2007.10129291.
13. Croft, W.B., Metzler, D., Strohman, T. (2009). Search engines-information retrieval in practice. In *Pearson education*. http://www.search-engines-book.com/.
14. Dooms, S., De Pessemier, T., & Martens, L. (2013). Movietweetings: A movie rating dataset collected from twitter. In *Workshop on crowdsourcing and human computation for recommender systems*, CrowdRec at RecSys 2013.
15. Felfernig, A., Boratto, L., Stettinger, M., & Tkalčič, M. (2018). Group recommender systems: An introduction. Springer Briefs in Electrical and Computer Engineering.
16. de Gemmis, M., Lops, P., Musto, C., Narducci, F., & Semeraro, G. Semantics-aware content-based recommender systems. In Ricci et al. [33] (pp. 119–159). https://doi.org/10.1007/978-1-4899-7637-6.
17. Goldberg, D., Nichols, D. A., Oki, B. M., & Terry, D. B. (1992). Using collaborative filtering to weave an information tapestry. *Communications of the ACM 35*(12), 61–70. https://doi.org/10.1145/138859.138867.
18. Harper, F. M., & Konstan, J. A. (2015). The movielens datasets: History and context. *ACM Transactions on Interactive Intelligent Systems 5*(4), 19:1–19:19. https://doi.org/10.1145/2827872.
19. Herlocker, J. L., Konstan, J. A., Terveen, L. G., & Riedl, J. T. (2004). Evaluating collaborative filtering recommender systems. *ACM Transactions on Interactive Intelligent Systems 22*(1), 5–53. https://doi.org/10.1145/963770.963772.
20. Kleinberg, J. M. (1999). Authoritative sources in a hyperlinked environment. *Journal of ACM 46*(5), 604–632. https://doi.org/10.1145/324133.324140.
21. Koren, Y. (2008) Factorization meets the neighborhood: A multifaceted collaborative filtering model. In *Proceedings of the 14th ACM SIGKDD international conference on knowledge discovery and data mining*, KDD 2008 (pp. 426–434). New York, NY, USA: ACM. https://doi.org/10.1145/1401890.1401944.
22. Lavrenko, V., & Croft, W. B. (2001) Relevance-based language models. In W. B. Croft, D. J. Harper, D. H. Kraft, & J. Zobel (eds.) *SIGIR 2001: Proceedings of the 24th annual international ACM SIGIR conference on research and development in information retrieval, September 9–13, 2001, New Orleans, Louisiana, USA* (pp. 120–127). ACM. https://doi.org/10.1145/383952.383972.
23. Liu, B. (2011). Web data mining: exploring hyperlinks, contents, and usage data. In *Data-centric systems and applications* (2nd ed.). Springer. https://doi.org/10.1007/978-3-642-19460-3.
24. Liu, T. (2009). Learning to rank for information retrieval. *Foundations and Trends in Information Retrieval 3*(3), 225–331. https://doi.org/10.1561/1500000016.
25. Lops, P., de Gemmis, M., & Semeraro, G. (2011). Content-based recommender systems: State of the art and trends. In F. Ricci, L. Rokach, B. Shapira, & P. B. Kantor (Eds.), *Recommender systems handbook* (pp. 73–105). US, Boston, MA: Springer.
26. Manouselis, N., Drachsler, H., Verbert, K., & Santos, O. C. (2014). *Recommender systems for technology enhanced learning: Research trends and applications*. Incorporated: Springer Publishing Company.
27. Mooney, R. J., Bennett, P. N., & Roy, L. (1998). Book recommending using text categorization with extracted information. In *Proceedings of the fifteenth national conference on artificial intelligence*, Madison, WI (AAAI 1998) (pp. 70–74).
28. Pazzani, M., & Billsus, D. (1997). Learning and revising user profiles: The identification of interesting web sites. *Machine Learning 27*(3), 313–331.

29. Pizzato, L., Rej, T., Chung, T., Koprinska, I., Kay, J. (2010) Recon: A reciprocal recommender for online dating. In *Proceedings of the fourth ACM conference on recommender systems*, RecSys 2010 (pp. 207–214). New York, NY, USA: ACM. https://doi.org/10.1145/1864708. 1864747.

30. Ponte, J. M., Croft, W. B. (1998). A language modeling approach to information retrieval. In W. B. Croft, A. Moffat, C. J. van Rijsbergen, R. Wilkinson, & J. Zobel (Eds.) *SIGIR 1998: Proceedings of the 21st annual international ACM SIGIR conference on research and development in information retrieval, August 24–28 1998, Melbourne, Australia* (pp. 275–281). ACM. https://doi.org/10.1145/290941.291008.

31. Rendle, S., Freudenthaler, C., Gantner, Z., & Schmidt-Thieme, L. (2012). BPR: bayesian personalized ranking from implicit feedback. CoRR http://arxiv.org/abs/1205.2618.

32. Resnick, P., Iacovou, N., Suchak, M., Bergstrom, P., & Riedl, J. (1994). Grouplens: An open architecture for collaborative filtering of netnews. In *Proceedings of the 1994 ACM conference on computer supported cooperative work*, CSCW 1994 (pp. 175–186). New York, NY, USA: ACM. https://doi.org/10.1145/192844.192905.

33. Ricci, F., Rokach, L., & Shapira, B. (Eds.) (2015). *Recommender systems handbook*. Springer. https://doi.org/10.1007/978-1-4899-7637-6.

34. Robertson, S. (2010). The probabilistic relevance framework: BM25 and Beyond. *Foundations and Trends in Information Retrieval 3*(4), 333–389. http://dx.doi.org/10.1561/1500000019.

35. Robertson, S. E. (1997) Readings in information retrieval. In *The probability ranking principle in IR* (pp. 281–286). San Francisco, CA, USA: Morgan Kaufmann Publishers Inc. http://dl. acm.org/citation.cfm?id=275537.275701.

36. Robertson, S. E., & Jones, K. S. (1988) *Relevance weighting of search terms* (pp. 143–160). London, UK: Taylor Graham Publishing.

37. Rocchio, J. J. (1971). Relevance feedback in information retrieval. In G. Salton (Ed.), *The smart retrieval system-experiments in automatic document processing* (pp. 313–323). Englewood Cliffs, NJ: Prentice-Hall.

38. Salton, G., Wong, A., & Yang, C. S. (1975). A vector space model for automatic indexing. *Communications of the ACM 18*(11), 613–620. http://dx.doi.org/10.1145/361219.361220.

39. Shaw, J. A., & Fox, E. A. (1994). Combination of multiple searches. In *TREC* (vol. Special Publication 500–225) (pp. 105–108). National Institute of Standards and Technology (NIST).

40. Singhal, A. (2001). Modern information retrieval: A brief overview. *IEEE Data Engineering Bulletin 24*(4), 35–43.

41. Sivic, J., & Zisserman, A. (2003) Video google: A text retrieval approach to object matching in videos. In *Proceedings of the ninth IEEE international conference on computer vision*, ICCV 2003 (Vol. 2, pp. 1470–1477). Washington, DC, USA: IEEE Computer Society. http://dl.acm. org/citation.cfm?id=946247.946751.

42. Xu, S., Bao, S., Fei, B., Su, Z., & Yu, Y. (2008). Exploring folksonomy for personalized search. In *Proceedings of the 31st annual international ACM SIGIR conference on Research and development in information retrieval*, SIGIR 2008 (pp. 155–162). New York, NY, USA: ACM. http://dx.doi.org/10.1145/1390334.1390363.

Chapter 6
Business Intelligence

Carl Anderson

Abstract Business intelligence is the combination of data, appropriate metrics, and the relevant skills, tools, and processes to make sense of what is happening in a business, and to make recommendations as to what should change or happen next. Most organizations attempt to leverage analytics to drive decision making. However, few of them are able to access the full value of what business intelligence has to offer. In this chapter, we will cover the most important skills, tools, and processes that will enable organizations to enhance their use of data and analytics. We cover types of analytical outputs or "levels of analytics" and how they relate to business intelligence and data science. We then detail four major types of analysis: descriptive, exploratory, inferential, and predictive. This is followed by a discussion on metric design, data dictionaries, and key performance indicators. We end by reviewing the role of business intelligence teams and some of the current trends in business intelligence.

6.1 Introduction

In order to respond to changing market conditions, all businesses need to make strategic, tactical, and operational decisions [13]. Should we increase our marketing spend? If so, in which channels, and targeting which segments? Which type of user is most likely to provide a referral? Which is our worst performing product, and should we cut it from our catalog? How many new call center support staff should we hire to support our growing demand? What should be our sales goals for next quarter? Did we meet our goals last quarter? If not, why? To be able to do all of these in an objective and data-driven manner requires business intelligence.

Business intelligence's (hereafter, BI) primay goal is to produce *actionable intelligence*. As such, that will tend to focus on aspects of the business that they can directly control—such as price, discounts, hiring/firing, inventory, product offerings, replenishment, and financial, marketing, and sales strategy—although it can also relate to

C. Anderson (✉)
Weight Watchers, New York, USA
e-mail: carl.anderson@weightwatchers.com

© Springer International Publishing AG, part of Springer Nature 2019
A. Said and V. Torra (eds.), *Data Science in Practice*, Studies in Big Data 46,
https://doi.org/10.1007/978-3-319-97556-6_6

assessing impact or risk of externalities—such as interest rates, competition, and weather.

Such intelligence requires the right combination of quality, relevant, and timely data (i.e., the raw ingredients for information and insight), the appropriate metrics (so as to track change), as well as the analytical skills, tools, and processes so that the business can not only make sense as to what is happening but adopt, rather than ignore, the recommended changes and actions.

Typically, BI attempts to answer questions such as:

- **Descriptive analytics**: What happened?
- **Diagnostic analytics**: Why did it happen?
- **Predictive analytics**: What will happen, if we do nothing and trends continue?
- **Prescriptive analytics**: What should we do? What will happen, if we do X?

For instance, imagine a large supermarket where a big snow storm is predicted to hit in 5 days, which is sufficient advance warning such that they could modify their restocking deliveries. Should they change their orders and if so, how? BI allows them to analyze their historical data and determine that in the last three big snowstorms, customers tended to stock up on milk, bread, batteries, flashlights and lanterns, and snow shovels. However, in those storms, milk and bread sold out early and thus the store did not capture all the demand, i.e. lost sales, and it also lead to unhappy customers who complained on social media, which is bad for the brand. Thus, maybe they should order more units this time.

Referring back to our five questions, using the milk example:

What happened?	Sold out of milk 12 h earlier than anticipated; Customers complained on social media
Why did it happen?	We did not meet customer demand
What will happen, if we do nothing and trends continue?	We will run out again and not maximize sales; We will see same pattern of customers complaining on social media
What should we do?	Increase regular milk order by 235%
What will happen, if we increase the milk order by 235%?	We predict that we will sell out just as storm hits, maximizing sales, minimizing spoilage (by not over-ordering), and minimizing complaints

In this case, it is clear how BI directly impacts the business, affecting top-line revenue (by capturing additional sales), bottom-line margins (by minimizing spoilage), and reducing risk to the brand (by minimizing complaints).

To answer those types of questions and provide those types of insights requires extracting information from the data. To achieve that, BI tends to determine and examine features such as:

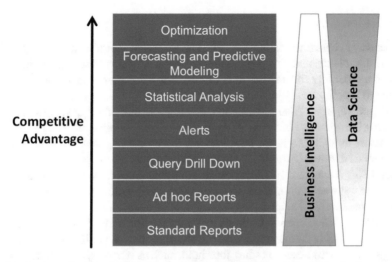

Fig. 6.1 A schematic of Davis's [7] "levels of analytics" and their relationship to business intelligence and data science. In general, BI is more associated with the lower descriptive and diagnostic levels while data science has greater association with the higher, more predictive and prescriptive levels. In this figure, Davis's two levels "forecasting" and "predictive modeling" are combined into a single level

- **Point estimates**: what were sales last month? What is the average session time on our website?
- **Comparisons among subsets of data**: how do sales patterns differ between this month and last? How does annual spends differ between men and women?
- **Association and correlation**: how does conversion percentage vary with user age?
- **Distribution**: what is the 95th percentile of shipping time? What is the median call waiting time?
- **Trends and forecasts**: how seasonal is our business? What is the overall trend in sales, controlling for seasonality? What are our predicted sales for Q3?

These facts, relationships, insights, and predictions feed into a number of outputs such as dashboards, reports, and analyses.

6.2 Levels of Analytics

To achieve the kind of intelligence illustrated above requires a set of different types of activities, coined "levels of analytics" by Davis [7], that try to address various types of questions (Fig. 6.1).

These levels are:

- **Standard reports**: a relatively fixed set of questions, metrics, and views (typically in tables, charts, and dashboards) that the business wants to see on a regular basis such as monthly financial reports or quarterly sales performance reports. These tend to report what happened and when.
- **Ad hoc reports**: "one-off" reports that answer individual questions that arise during the course of business. They are typically descriptive reports, requested by business users, that address questions such as how many, how often, or where?
- **Query drill down** (or online analytical processing, OLAP): more analytical activities diving down into underlying causes for a phenomenon. Why did we see a spike in sales last week? What is driving the increase in ship time to our customers? This is where individual analysts may spend a lot of their time.
- **Alerts**: proactive communications that alert the business of some change in key metrics allowing them to make some operational change. For instance, alerts might trigger if a metric exceeds some threshold, such as if average call waiting time at a call center exceeds 10 min or if inventory for product X has dropped below Y units. This is the basis of "management by exception," a best practice that brings issues to the attention of business users when they deviate from some expectation or norm.
- **Statistical analysis**: objective but probabilistic analysis that attempts to quantify signal versus noise. Is this really a trend or just random variation? How much of this pattern is driven by seasonality? This would be tackled by analysts and data scientists, such as for analyzing results of A/B tests (discussed in Sect. 6.3.5 in this chapter).
- **Forecasting and predictive modeling**: models that analyze historical data to predict what is likely to happen. What if these trends continue and how will it affect my business? How much is needed? When will it be needed? For instance, retailers can predict demand for products from store to store and casinos may predict which VIP customers will be more interested in particular vacation packages. This would be tackled by analysts and data scientists in a variety of departments, such as supply chain, marketing, or sales.
- **Optimization**: more complex models that attempt to maximize some objective, such as sales, retention, or time. How do we do things better? What is the best decision given that this is a multidimensional problem? For instance, what is the best way to optimize IT infrastructure given multiple, conflicting business and resource constraints? This would typically be the domain of data scientists or other specializations such operations research.

These levels range from reporting which is *backwards-looking* (what happened?), alerts (what is currently happening?), analysis (why is this happening?; what should we do?) to more *forward-looking* predictions and optimizations (what will happen if we do X?; what is our optimal strategy?). See Table 6.1.

Table 6.1 Aligning levels of analytics with types of questions and a temporal perspective: past, present, and future

Pespective	Past	Present	Future
Questions	What happened?	What is happening?	What will happen?
	Why did this happen?	Why is this happening?	What should we do?
Activities	Ad hoc reports	Alerts	Statistical analysis
	Query drill down	Query drill down	Forecasting
	Standard reports		Optimization

All of these are important. However, many organizations get stuck on the backwards-looking components, churning out dashboards and reports without a clear reason and without being tied to a particular business decision. More data-driven organizations, though, are likely to be involved in true analysis (will attempt to answer why) and the more forward-looking predictive and prescriptive activities. They provide concrete recommendations of what to do, why, and what is the expected result [1].

6.2.1 Relationship to Data Science

What is the relationship between BI and data science? In practice, there is significant overlap between the two. BI tends to focus more on the backwards-looking components, such as scheduled reports, ad hoc reports, and query drill downs. Data science tends to focus on the more forward-looking components such as predictive models and optimization. However, there are certainly common methods and, in industry, blurry roles with many BI analysts doing data science, and many data scientists doing BI.

For instance, BI, like data science, certainly can cover regression and other data mining tools such as clustering and tree-based models (covered in Chap. 3). On the other hand, data science, like BI, typically needs to start with descriptive and exploratory data analysis (both detailed below) to understand the scale, scope, and quality of datasets that they plan to use. Further, some specialties such as text analytics and sentiment analysis, which were once domain of data scientists have become more democratized and are now more likely to be found in BI.

Overall, there are common types of data, tooling, and methodologies that both use. For instance, both BI analysts and data scientists often require familiarity with databases and query languages such as SQL. However, data scientists *tend* (with many exceptions) to be more self-sufficient and more "full stack" than BI analysts in terms of coding skills, often possessing skills ranging from extracting data from sources (whether it be web-scraping, a database, or an application programming interface [API]), loading it in some data store (such as SQL, NoSQL, hadoop, or a filestore), developing models, data visualization, and building out some frontend interface or API to deliver the results. While it is possible to be a productive business intelligence analyst who does not code, that is far less likely to be the case with a data scientist. Data scientists tend to be stronger in algorithms, statistics and machine learning, and linear algebra.

6.3 Types of Analysis

Given the levels of analytics, what are the types of analysis and how is it actually performed? Leek claims that there are six types of analysis (see [1]), although we will focus primarily on the first four, which are the most common in a BI and data science business context:

- **Descriptive Analysis**
- **Exploratory Data Analysis**
- **Inferential**
- **Predictive**
- Causal
- Mechanistic.

6.3.1 Descriptive Analysis

The most basic analysis is descriptive. This summarizes a dataset quantitatively, and rather than predict attributes of the population from which the data came, it simply characterizes the data at hand. Descriptive analysis, especially counts and means, forms the basis of many reports, alerts, and dashboards, a core component of BI.

In many cases, descriptive analysis summarizes a single variable, that is, univariate distributions. This means generating summary metrics that map a set of raw values into a few individual values that cover attributes of the distribution such as sample size, location, spread, and shape.

Location metrics include:

Mean	Arithmetic mean
Median	50th percentile
Mode	Most frequently occurring value
Geometric mean	Average when there are multiplicative effects at play, such as compound interest
Harmonic mean	Arithmetic mean of the reciprocals, typically used when dealing with rates, such as velocity

Dispersion metrics include:

Minimum	Smallest value (0th percentile)
Q1	25th percentile
Q3	75th percentile
Maximum	Largest value (100th percentile)
Interquartile range	Q3–Q1
Range	Maximum–minimum
Standard deviation	Dispersion about the arithmetic mean
Variance	Square of standard deviation
Standard error	Standard deviation divided by square root of sample size
Gini coefficient	Typically a measure of the inequality of income among individuals within a population but can be used more broadly.

Shape measures include:

Skew	measure of the asymmetry of a distribution. If it has a larger tail to the right, that is towards larger value, it has positive skew. For instance, time spent on website per session or house prices are positively skewed
Kurtosis	measure of the sharpness or flatness of a distribution. Platykurtic distributions tend to have "fatter" tails while leptokurtic distributions have thinner tails and stronger, shaper peaks.

Many variables of course are categorical. For these, simple summary tables showing raw counts, or normalized metrics such as percentages, are a very common and yet powerful component of BI. Further, to understand the relationship among categorical variables, one might cross-tabulate counts for a contingency table such as Table 6.2 which exhibits an association among gender and product preference.

6.3.2 Exploratory Data Analysis

Descriptive analysis while simple, intuitive, and easy to implement can be dangerous if that is all that one considers. Reducing a sample or distribution to a few summary numbers can be very misleading and does not tell the whole story. For instance, a bimodal distribution may have the same mean as a unimodal distribution. Varying

Table 6.2 An example cross tabulation of two categorical variables, gender and product preference, illustrating an association between the two

Gender	Preferred product A (%)	Preferred product B (%)
Male	76	24
Female	43	57

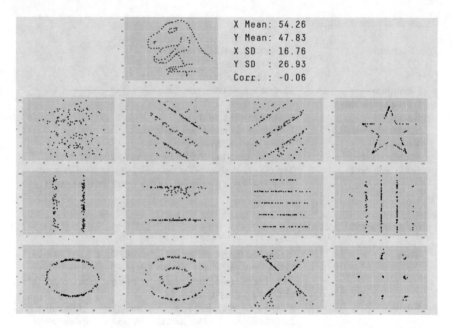

Fig. 6.2 Matejka and Fitzmaurice's [12] "datasaur dozen" illustrating that different bivariate arrangements or distributions (13 scatterplot panels above) can have the same summary metrics (metric panel at top right). (Reproduced with permission from https://www.autodeskresearch.com/publications/samestats)

degrees of kurtosis might result in the same mean value. That is, multiple distributions can give rise to the same summary metric values.

Anscombe [4] illustrated this point well by generating a carefully-crafted fake dataset, his well-known "Anscombe's quartet," of four widely different bivariate samples in variables x and y that had the same sample size, means, variances, correlations, and regression equations. More recently, Majejka and Fitzmaurice [12] used simulated annealing to generate their "datasaur dozen" making the same point. Their panels in Fig. 6.2, including the dinosaur, all have the same mean(x), mean(y), sd(x), sd(y), and corr(x, y) and yet on inspection we can see that the samples vary significantly.

The point here is that analysts have to study, inspect, and plot the data, a process termed exploratory data analysis (EDA) by Tukey [18]. Summary metrics alone will not suffice. While this seems obvious, it is often overlooked.

There are several reasons to perform EDA and plot the data, detailed in the following sections.

6.3.2.1   Understand Shape

Suppose that the following descriptive statistics of sample of house sales prices [8] were provided: $\mathtt{min} = \$12k$, $\mathtt{median} = \$160k$, $\mathtt{max} = \$755k$, $\mathtt{skew} = 1.74$, and $\mathtt{kurtosis} = 5.1$. It is hard to imagine the shape. While the $\mathtt{median} - \mathtt{min}$, $\mathtt{max} - \mathtt{median}$, and $\mathtt{skew}$ might give you a sense of the positive skew, the best way to inspect the data is visually (Fig. 6.3). A picture says a thousand words.

6.3.2.2   Check Data Quality

Another reason to plot data is to check data quality. Outliers, duplicates, and missing data are often obvious when plotting the data with a histogram, scatterplot, or boxplot. See Fig. 6.4. Our eyes and brains are incredible pattern detectors and we can often pick our anomalies which are hard to define metrics for. Moreover, it might be easier, in fact, to spot bad data than imagine good data.

6.3.2.3   Check High-Level Trends

When plotting data, one can more easily spot high-level trends, check the general ranges in numbers, and see whether they match one's intuition. Figure 6.5 shows the price of Bitcoin (in US dollars) during its meteoric rise during 2017. If an analyst had

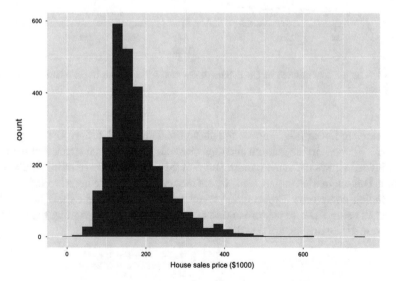

Fig. 6.3 House sales price from the Ames dataset [8]. Plotting the histogram of the data provides a richer way to understand the distribution, such as the degree of skewness, than summary metrics alone

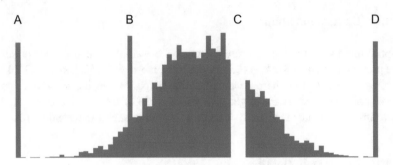

Fig. 6.4 Example of the sorts of errors and features that can be detected with a simple histogram: **a** default values, such as 1, 0, or 1900-01-01; **b** incorrectly entered or duplicate data; **c** missing data; and **d** default values, such as 999. Reproduced from [1]

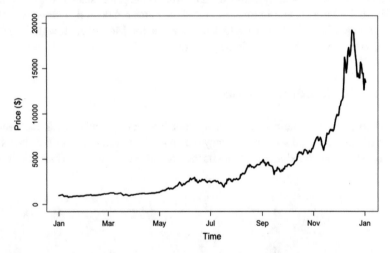

Fig. 6.5 The price of Bitcoin in US dollars during 2017. Data from https://charts.bitcoin.com/chart/price

some expectation or prior knowledge about the data, say that they knew it reached around $20k/coin in the autumn and then crashed, then they can check that the chart does indeed shows the same pattern and that they did not receive other similar data such as Bitcoin market cap instead (where the max would have been around $321B). Developing some prior expectations about what the data might show before they inspect it, using EDA to explore and to confirm, and then questioning the situation when something differs, are one the hallmarks of a good analyst.

6.3.2.4   Generate Hypotheses

The previous three uses cases for EDA cover getting an overall sense of the data, such as what variables are present, what distributions or associations do they have, as well as checking data quality. However, BI can use EDA to generate hypotheses and explore opportunities. By highlighting trends, associations, and other patterns, BI analysts can then question what might be causing the patterns that they see, bringing us to the second question that we outlined in the introduction (Why did it happen?).

By focussing on variables that the business can control, they can suggest experiments to find out more or suggest changes. For instance, suppose that a chart of number of ad conversions (i.e. number of people who bought an item from an ad) versus advertising spend for a certain channel (such as Facebook) shows ad saturation and diminishing returns. An analyst might realize that spending more money on the channel is futile and the business might do better by diverting some of that money to a different, unsaturated channel (such as TV or Twitter), or to a different segment (subset of potential customers).

6.3.3   Inferential Analysis

Descriptive analysis and EDA are both aspects of *descriptive* statistics. That is, they describe properties of a sample. The other major branch of statistics is *inferential* statistics. Here, one infers the properties of the source of the data—the population—and one can use its probabilistic framework to answer questions or make inferences about different samples. Importantly, inferential statistics provides a much more objective way to analyze differences, associations, and trends, and provides a much stronger context with which the business can make a decision.

For instance, suppose that in a monthly subscription business, one set of people were given an initial discount or "teaser" rate for 3 months, and then switched to the regular rate, while another set of people had to pay the regular rate from day 1. If the churn rate at the end of months 1 and 2 are different, say those with the higher rate have a 2% higher churn rate, how likely is it that this is noise versus some real difference among the groups? Rather than simply report the difference (descriptive analysis), it is much more powerful to a decision maker to report that while the non-teaser group had a 2% higher churn rate, there is a 56% chance we would have seen those results by chance alone (if the null hypothesis were true). From this, the business might conclude that if there is no underlying difference in the churn rate, and given than the teaser rates costs the company money in terms of lower revenue, they might get rid of the discount.

The types of questions that one can answer with statistical inference include (but are not limited to):

- **Standard error, confidence intervals, and margin of error**: how much confidence can one have in this particular sample mean or sample proportion? How much might it vary due to chance if one were to repeat the experiment?
- **Expected mean of single sample**: is this sample mean significantly different from the expected value?
- **Difference of means from two sample**: are the means from these two samples significantly different from each other? More technically: what is the likelihood that we would observe this difference of means, or greater, if the null hypothesis were true that there is no difference in the population means for the two samples?
- **Sample size calculations and power analysis**: what is the minimum sample size needed, given what is currently known about the process, to achieve a certain level of confidence in the data? These types of statistical tools are important for planning A/B testing.
- **Distributions**: is the distribution of values in this sample consistent with a normal (bell-shaped) distribution? Do these two samples likely have the same underlying distribution?
- **Regression**: suppose that a well-designed experiment is conducted in which one (independent) variable is controlled systematically while controlling as much as possible for all others factors, and then regression is fitted. How much confidence can we have in that trend line? How much is it likely to vary (in both gradient and intercept) if we were to repeat the experiment many times?
- **Goodness of fit and association**: given a categorical variable (say, product category), do the frequencies or counts (say, purchases) match a set of expected relative frequencies? Is there a relationship among two variables, one of which is categorical?

6.3.4 Predictive Analysis

It is within predictive analysis that there is the greatest overlap with data science. Some predictive models are very common in BI. For instance, retail organizations will need to forecast demand, primarily to ensure that there is sufficient inventory to sell but also for warehousing, customer support, and setting appropriate company goals. In addition, most organizations will need to forecast financials. Often these forecasts involve simple regressions or time series analysis, such as as ARIMA or exponential smoothing [11].

Many products and services exhibit strong seasonality due to weather (say, beach towel sales), religious holidays (Christmas or Hanukkah decorations), or government deadlines (tax season). Many websites exhibit very strong weekly seasonality with traffic patterns on the weekend very different from weekdays. For all these, one might decompose the data (say using LOESS [6]) to separate out the seasonal component from the general trend. This then provides a clearer view into the business.

Many businesses will also perform some sort of segmentation of their customers. This is especially true of marketing teams who want to understand who the customers are, how they find "look alikes" (other customers similar to current customers), and how best to target them. This too falls squarely within BI.

However, there are many other predictive models that fall more within the realm of data science (reviewed by [16]; see also Chaps. 2, 3, 5, and 6 of this book). For instance, predicting who to date (dating apps), which movie to watch or shirt to buy (recommender systems), and who to approve for a loan (banking).

6.3.5 Causal and Mechanistic Analysis

While descriptive, exploratory, inferential, and predictive comprise most of BI, for completeness, there two additional types of analysis, which we will not cover in detail.

Causal analysis refers to situations with controlled experimental conditions, intended to isolate individuals factors or effects. The most common cases in business are those of A/B testing (reviewed in [1, 17]) in which experiments are run to optimize website metrics, such as sign ups, check outs, or other engagement. Imagine a website has a button with text "sign up" and that the business hypothesizes that different messaging such as "learn more" might increase the click through rate. They create two versions of the website, one set of visitors get to see "sign up" and the other visitors get to see "learn more". After a week or two, they analyze which version had a higher rate. If "learn more" exhibits a statistically significant higher rate of clicks, then the company will roll out "learn more" messaging to all visitors.

Mechanistic analysis is associated less with BI and more with fundamental scientific research and development, and the term "modeling" more than "analysis." Mechanistic modeling and analysis represents a very deep understanding of a system, which comes from studying a stable system in a very controlled manner with many experiments over many years—hence the association with basic science. This situation does not tend to occur within most businesses, with some exceptions, such as R&D departments in pharmaceuticals and engineering.

6.4 Metric Design

The last section discussed types of analysis. Just as important is deciding which data one should track or analyze. Choosing an appropriate set of metrics with the right characteristics is crucial. They effectively become the lens through which the executives, staff, and investors view the business. If they are poor metrics which are biased, indirect, or they do not reflect the goals and strategy of the leadership, the business can optimize for the wrong thing or otherwise be lead astray and fail. As such, metric design and selection is a key responsibility of BI.

BI can consist of some simple metrics such as counts and averages of raw data: the number of users who churned last month, the average basket size of an order, or the highest number of contracts won by a salesperson. However, many metrics are *derived* metrics, a combination of two or more other values. Bounce rate is the *proportion* of website visitors that leave the site after visiting a single page (i.e., # that leave / # total). Net revenue is total sales *minus* returns, allowances, and discounts. These rates, differences, ratios, and other more complex metrics need to be designed to have the appropriate properties. If they do not, then they might cause some confusion or, more importantly, they might give an inaccurate or biased view of the business.

6.4.1 Metric Traits

When choosing or designing a metric, it should posses the following traits:

- Simple
- Standardized
- Accurate
- Precise
- Robust
- Direct.

6.4.1.1 Simple

Do not make metrics unnecessarily complex; that is, design them to be "as simple as it can be, but not simpler" (Einstein). A metric such as "Sales(Units)" is much more easily understood if it is simply the "numbers of units sold within some time frame" than say "number of units sold within some time frame, excepting those returned or exchanged with 45 days, or which were found to be fraudulent transactions, or which. . . "

There are several benefits to such simplicity. First, simple metrics are easier to define and explain to a colleague; thus, there is less chance of confusion and staff are more likely to use the information for their decision making. Second, simple metrics are easier to implement. This means that there is less chance of errors and also is easier to maintain and document. Third, simple metrics are more likely to be comparable to other teams or organizations.

6.4.1.2 Standardized

Standard metrics should be adopted unless there are very good reasons to deviate from them. If a standard definition is used, such as for bounce rates, net revenue, or sessions, then it becomes possible to compare one's performance with other teams, competitors, or other industry benchmarks. For instance, for "active users," Facebook only considers users that are logged in, whereas Yelp considers both logged in users as well as guests. As such, they are not comparable.

Using standards will help avoid confusion. It will also make it easier for staff new to a team or organization to get up to speed. The key, however, is to make sure that the definition is standardized across the whole organization. While it sounds obvious, it is much more common than one would expect for different teams to deviate from the norm and have their own variants. Two different dashboards might appear to show the same metric but if they have different underlying logic, confusion will reign.

6.4.1.3   Accurate

One should strive for metrics that are accurate. Accuracy refers to a metric's ability to track the true mean value. If you compare to archery, it is the equivalent to being on target (Fig. 6.6). "Sales(Units)" accurately captures demand and the number of units that customers checked out in their baskets. Other metrics, however, might exhibit some bias. For instance, in surveys, there is often response bias in which especially happy or unhappy customers, or other segments such as men versus women, are more likely to complete a survey providing a set of responses that are unrepresentative of the whole population.

An example of bias comes from Coca Cola's testing and introduction of New Coke in the 1980s [19]. They ran user tests that showed very positive results, even

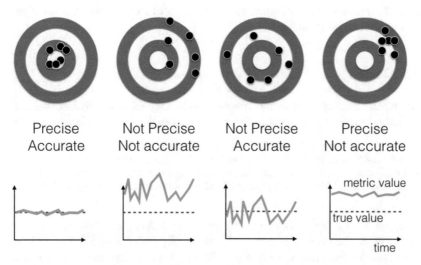

| Precise | Not Precise | Not Precise | Precise |
| Accurate | Not accurate | Accurate | Not accurate |

Fig. 6.6 Precision (being stable or clustered) and accuracy (being on target) illustrated with two-dimensional data. Inaccurate metrics have a bias such that their mean differs from the true mean in a systematic way. Precision captures the variability: how much the mean value would differ if one were to repeat the experiment multiple times and collect new samples of the same size. (Reproduced from [1])

compared to regular Coke. When it launched, however, the product tanked. Why? Users found it too sweet. The problem is that testing involved "sip tests," and we tolerate greater sweetness in sips than in gulps. Had they tested a "normal" user experience (gulping a mouthful on a hot day), they would have been able to optimize for the real-world taste, experience, and preference.

One should try to understand the processes generating the data and any biases that it might contain. Sometimes there is no choice other than use a biased metric but it might be possible to do an analysis and determine some correction factor. If a correction factor is used, bear in mind that conditions change and that it may need to recalibrated periodically.

6.4.1.4 Precise

Metrics should be precise. That is, they should return similar values if repeated with the same conditions. Returning to the archery analogy, it is the equivalent of a set of arrows being close to each other (Fig. 6.6).

One tool or lever to control precision is sample size. The larger the sample size, the smaller is the standard error. The relationship is not linear, however. Because standard error of the mean equals the standard deviation divided by the square root of the sample size, to halve the standard error, we need to quadruple the sample size [9].

Unless there is some validated reference data, we are not always going to know whether our metrics are inaccurate. However, we will likely soon discover if our metrics are imprecise (unstable).

6.4.1.5 Robust

Define metrics that are statistically "robust," that is, they are relatively insensitive to individual extremal values. Consider this example from the *San Francisco Chronicle*:

> The average tech wage on the mid-Peninsula (San Mateo County) last year was $291,497...
> A likely explanation for the distortion: Facebook chief executive Mark Zuckerberg took only a dollar in salary but reaped $3.3 billion exercising Facebook stock options in 2013...
> if you remove $3.3 billion from the total, you end up with an average tech wage of roughly $210,000.

Using an average here is inappropriate given the high degree of positive skew of the salary data. The average is increased significantly (more than 35%) by a single outlier value. In this case, a median is a far better choice because it is robust to extremal outliers and better captures the location or middle of the data.

It should be noted that there are times when one wants to design a metric that is specifically sensitive to boundary values. For instance, peak load on a website is meant to capture the rare maximal values encountered.

It is possible to estimate or visualize robustness by resampling. Take a subset of the data and compute the metric. Repeat many times (with replacement) to get a set of metric values and plot their distribution.

6.4.1.6   Direct

Try to define or use metrics that directly measure the underlying process. `Sales (units)` is direct: 1 unit checked out in basket = 1 unit in the metric. However, consider student performance in a school. We do not have a good, direct performance measure and so educators and governments often rely on the proxy of test scores. The number of phone calls to a customer service line might be a proxy for customer happiness. Thus, proxy metrics are *indirect* measures.

Proxies are not necessarily correlated 100% with the underlying process. They may also be more easily subject to "gaming the system" (say, bad teachers teaching just what will appear on the test), subject to external influences that affect the proxy but not the underlying process, and subject to biases.

Proxies are used when it can be hard or impossible to measure the underlying process. However, they might also be used because that is the only data available. In the example of customer happiness, while number of phone numbers might be quick and easy to compute, a better and more direct measure—even though it requires more effort—would be to survey customers and ask them directly, and translate their responses to a standard metric such as net promoter score [15].

6.4.2   Absolute Versus Relative

One last, but important, aspect of metric design is to consider whether to use absolute or relative metrics as they can paint a very different picture. Imagine a business in which 25% of their members are designated as VIPs. They run a successful campaign to grow the number of new members. Assuming that that none of these are VIPs, the absolute number of VIPs is unchanged and the number of non-VIPs has increased. However, suppose that a dashboard used the metric of the proportion of members that are VIPs. That value will have decreased. An executive watching the numbers but unaware of the campaign might be alarmed that behavior of VIPs has changed—perhaps they are churning more or spending less, she worries. Conversely, suppose that the campaign was to focus on increasing repeat customers. Here, the total number of members might not change but the proportion of VIPs might increase.

Table 6.3 Example of a data dictionary

Field	Description	Type	Example
MemberId	The unique ID of the member	integer	1463463
FirstName	First name of the member	varchar(255)	John
LastName	Family name of the member	varchar(255)	Smith
JoinDate	Date that member joined	date	2018-01-03
MemberLevel	Type of member: one of regular or VIP	string	VIP

6.5   Data Dictionary

We mentioned earlier that using simple and standard metrics will result in less confusion among staff. However, a well-run BI team will also publish a data dictionary.
This is a glossary, a list of all the key business terms and metrics with clear, unambiguous definitions and examples. An example data dictionary is shown in Table 6.3.

At WeWork (a global provider of flexible, physical workspaces), prospective
members check out facilities by signing up for a tour. Importantly, some people
may tour different locations, or come back for a second tour to show other members
of their organization before signing off on their new space. While their various dashboards had a metric called "tours," that term did not align across teams. The process
of creating a data dictionary fleshed out two different metrics [3]:

- `Tours completed–Volume` captures the absolute number of tours taken,
  which our Community team, who staff such tours, monitor.
- `Tours completed–People` captures the unique number of people who signed
  up for a tour. This can then feed into a lead conversion metric, which our sales and
  marketing teams track.

To get to those definitions required lots of discussions and alignment among
teams. However, once there is agreement, the data dictionary can be created and
published where all staff can access it (such as an internal wiki). When the BI team
implement those metrics in dashboards and reports matching those definitions, it can
be incredibly powerful—so much so that Randall Grossman (Chief Data Officer for
Fulton Financial) says "A trusted data dictionary is the most important thing that a
[Chief Data Officer] can provide to business users" (pers. comm.) [2].

6.6 KPIs

Having a set of good metrics, a set of dashboards to report them over time, and a data dictionary, will afford some level of business intelligence. They might show what is happening but offer no guidance whether any changes are good or bad, or whether this is what the business needs to achieve. It is for this reason that businesses often set goals using a set of KPIs.

A KPI, or key performance indicator (sometimes known as Key Success Indicators) is a high-level metric tied to some strategic goal of the company. Examples might include "increase online conversion by 5% in Q3," "increase brand awareness by 20% by year end," or "decrease capital expenditure per square foot by $100."

KPIs need to exhibit a number of traits:

- **Clearly defined**: they need to associate a goal, a time frame, and a metric, all of which are clear and unambiguous.
- **Measurable**: they need to be quantifiable, i.e. generate a number that can be tracked over time, and ideally be direct and not a proxy. If you cannot measure it, you cannot track it.
- **Have targets**: a target of "increase annual revenue" is insufficient for a KPI. The business does not want a sales team stopping work as soon as revenue is one Euro more than last year. The goal must be specific, a stretch goal, and also achievable. Either too low (easily achieved) or too high (unachievable), and people will give up.
- **Visible**: a set of KPIs are often called a strategic dashboard. Those metrics should be clearly visible and accessible to all the staff, such as on a dashboard on a monitor. These metrics should be on the top of everyone's mind so that people can assess performance and know what is and is not working.
- **Reflect what the organization is trying to achieve**: do not fall into the trap of just tracking what is easily measurable. If the goal is to improve customer satisfaction, then design a metric that reflects that—likely direct questions through customers surveys—and not using some proxy or inaccurate metrics that your BI system happens to have available.

In short, KPIs should be SMART: Specific, Measurable, Achievable, Result-oriented, and Time-bound [10].

6.7 Trends and Tools

What are some of the trends and tools in BI? First off, despite the significant developments in analytical tools, Microsoft's spreadsheet application, Excel, is still by far the most common BI tool in use today. It is a tool that is relatively easy to grasp and yet has significant power user functionality. It is also accessible in terms of price and ability to install to even the smallest business. The downside is that data and models tend to reside in files on individuals' computers, or in forgotten email threads, rather than on a sanctioned, shared, and versioned space. As such Excel data is often "dark data."

In small organizations, individual analysts may have to perform a number of BI tasks themselves such as data gathering, cleaning, and visualization in order to perform business intelligence. In larger organizations, however, dedicated BI teams will support the rest of the staff. A BI team's responsibilities are generally broad:

1. vet new data sources.
2. work with the data engineering team to define the data warehouse data models such as database views, dimension, and fact tables that will make downstream analysis easier.
3. work with the business to define the key metrics and then implement them in the databases and/or BI tools.
4. either the BI or data engineering team will implement ETLs (Extract, Transform, Load) which are processes to consume and transform data, such as data warehouse tables with KPIs that are often rebuilt or refreshed each night.
5. implement data quality processes to alert their team if individual metrics are going out of range. (This is in addition to data quality checks on the raw data landed in the database.)
6. help define, implement, and publish a data dictionary.
7. determine appropriate access to data.
8. provide business intelligence tools (such as Tableau, Looker, Domo, QlikView, Spotfire, Pentaho, and Alteryx) to those that need it and provide general support and training.
9. in some companies, dashboarding and report generation are centralized in a BI team, i.e. the BI team create all the reports, while in other companies, it is more distributed to analysts across other teams, especially if the BI tool is "self-service."
10. be a general resource for BI and data visualization best practices.
11. keep up to date with latest trends and vet new BI technologies.

At the enterprise level, where a BI team supports a set of users ideally with clean, high-quality, and curated data, a data dictionary, and a "single source of truth" (a central location to get data such that all staff are using the same data) there has been significant innovation over the last decade. Such tools are increasingly becoming democratized. These tools, which were once only in the hands of executives and the formal analysts, are increasingly getting out to non-analysts at the front lines of the organization who become empowered to make real time operational decisions. The BI tools are becoming cheaper, more intuitive, and easier to use. They offer more types of visualizations, can hook into other libraries such as D3 (https://d3js.org/), and provide APIs so that the metrics can be consumed in more ways. BI is increasingly "mobile first" so that tools are designed to work well on mobile devices such as phones and tablets. They are also becoming integrated into more channels. For instance, Looker, a BI tool vendor, provides a slackbot (a tool within the messaging and collaboration platform Slack) in which analysts, decision makers and other operational staff can bring up key charts or provide answers by asking questions using simple natural language.

Increasingly, customers interact with businesses via mobile, such as smartphone apps that have access to the phone's GPS, and so businesses have ever richer data about where customers are, where they go, and when. Consequently, location intelligence is becoming a powerful component of BI, allowing businesses to analyze spatial temporal patterns, optimize products, pricing, marketing campaigns, and even geo-target individual customers.

The last decade has also seen significant developments in database technologies. There are now many more, and more performant, databases and technologies tailored for unstructured data (MongoDB, PostgreSQL), big data (Hadoop, Cassandra), graphs (Neo4j, Aster), or time series (Influx, Riak) which in turn drive innovation in the tools used to visualize such data at scale or in near real time. Advances in data science, such as natural language processing, mean that BI analysts can now access processed stream of customer-generated or other text and analyze previously out of reach metrics such as sentiment. While there are many tools to alert users for certain types of patterns, such as anomalies from fraud or spikes in server load, we are beginning to see innovation in automated machine learning (AutoML), where interesting trends, patterns, or charts are discovered by algorithms and surfaced for an analyst to review. An example is the explore tab in Google sheets [14]. Thus, data science functionality and products are increasingly become democratized to BI.

6.8   Conclusion

Business intelligence is an important staple of almost all businesses. While organizations might differ greatly in terms of the volume of their data, the degree of data quality and centralization, or the sophistication of their tooling, the fact is that executives and other decision-makers rely on and demand BI, using at least some basic metrics about the business to make strategic, tactical, and operational decisions. BI provides a way to track the pulse of a business, to surface the key metrics, trends, and changes, and to track progress against goals.

The BI community has responded to such demands and there is a rich set of tools, both open source and proprietary, that provide powerful exploratory data analysis, dashboarding, and ways to execute ad hoc investigative queries.

There is also an increasing awareness that being data-driven provides a competitive advantage. One report, controlling for other factors, found that data-driven organizations have a 5–6% greater output and productivity than their less data-driven counterparts [5]. Business intelligence is at the heart of being data-driven. It provides a way to surface what is happening, and to help decisions makers. However, it can also be a source of innovation itself. As Todd Holloway, former head of data science at Trulia, remarked (pers. comm.), "the best ideas come from the guys closest to the data...they often come up with the good product ideas." Business intelligence is the window to the data.

Acknowledgements I would like to thank Carlo Bailey, Kashuo Bennett, Andrew Heumann, Brian Holland, and Anne Moxie for their thoughtful comments and feedback on an earlier version of this manuscript.

References

1. Anderson, C. (2015). *Creating a data-driven organization*. Sebastopol, CA: O'Reilly.
2. Anderson, C. (2018). Data dictionary: a how to and best practices. https://medium.com/@leapingllamas/data-dictionary-a-how-to-and-best-practices-a09a685dcd61
3. Anderson, C., & Li, M. (2017, June 23). Five building blocks of a data-driven culture. *TechCrunch*. https://techcrunch.com/2017/06/23/five-building-blocks-of-a-data-driven-culture/.
4. Anscombe, F. J. (1973). Graphs in statistical analysis. *American Statistician, 27*, 17–21.
5. Brynjolfsson, E., Hitt, L. M., & Kim, H. H. (2011). Strength in numbers: How does data-driven decisionmaking affect firm performance? *Social Science Research Network*. https://papers.ssrn.com/sol3/papers.cfm?abstract_id=1819486.
6. Cleveland, R. B., Cleveland, W. S., McRae, J. E., & Terpenning, I. (1990). STL: A seasonal-trend decomposition procedure based on Loess. *Journal of Official Statistics, 6*(1), 3–73.
7. Davis, J. (2009). Eight levels of analytics (p. 4). Cary, NC: SAS Institute Inc.
8. De Cock, D. (2011). Ames, iowa: Alternative to the boston housing data as an end of semester regression project. *Journal of Statistics Education, 19*(3). https://doi.org/10.1080/10691898.2011.11889627.
9. Diez, D. M., Barr, C. D., & Çetinkaya-Rundel, M. (2015). *OpenIntro statistics* (3rd ed.). https://www.openintro.org/stat/textbook.php.
10. Doran, G. T. (1981). There's a S.M.A.R.T. way to write management's goals and objectives. Management review. *AMA FORUM, 70*(11), 35–36.
11. Hyndman, R. J., & Athanasopoulos, G. (2013). *Forecasting: Principles and practice*. https://www.otexts.org/fpp.
12. Matjeka, J., & Fitzmaurice, G. (2017). Same stats, different graphs: Generating datasets with varied appearance and identical statistics through simulated annealing. In *Proceedings of the 2017 CHI Conference on Human Factors in Computing Systems, Denver, Colorado, May 6–11* (pp. 1290–1294).
13. Moxie, A. (2012). Guidebook: Measuring the half life of data. Nucleus Research, Report M36. https://nucleusresearch.com/research/single/guidebook-measuring-the-half-life-of-data/.
14. Ranjan, R. (2016). *Explore in Docs, Sheets and Slides makes work a breeze and makes you look good, too*. https://docs.googleblog.com/2016/09/ExploreinDocsSheetsSlides.html.
15. Reichheld, F. F. (2003, December). One number you need to grow. *Harvard Business Review*.
16. Siegel, E. (2013). *Predictive analytics*. Hoboken: Wiley.
17. Siroker, D., & Koomen, P. (2013). *A/B testing*. Hoboken: Wiley.
18. Tukey, J. W. (1977). *Exploratory data analysis*. Pearson.
19. Webber, S. (2006, May/June). Managements great addiction. *Across the Board*, 43–48.

Part III
Tools

Chapter 7
Data Privacy

Vicenç Torra, Guillermo Navarro-Arribas and Klara Stokes

Abstract In this chapter we present an overview of the topic data privacy. We review privacy models and measures of disclosure risk. These models and measures provide computational definitions of what privacy means, and of how to evaluate the privacy level of a data set. Then, we give a summary of data protection mechanisms. We provide a classification of these methods according to three dimensions: whose privacy is being sought, the computations to be done, and the number of data sources. Finally, we describe masking methods. Such methods are the data protection mechanisms used for databases when the data use is undefined and the protected database is required to be useful for several data uses. We also provide a definition of information loss (or data utility) for this type of data protection mechanism. The chapter finishes with a summary.

7.1 Introduction

The amount of information gathered by organizations, both in the public and the private sector, makes privacy at stake. Organizations keep track of all products and services we have bought, places we have visited, and where we are at any time. Databases also store information of our finances and even public offenses.

Although human rights and international legislation establish that privacy is a fundamental right, the application of this right is in tension with our interest in personalized services, in business and government interest in big data to increase profits and planning (e.g., city planning), and in public interest for high levels of

V. Torra (✉) · K. Stokes
University of Skövde, Skövde, Sweden
e-mail: vtorra@his.se; vtorra@ieee.org

K. Stokes
e-mail: klara.stokes@his.se

G. Navarro-Arribas
Universitat Autònoma de Barcelona, Campus UAB, Bellaterra, Catalonia, Spain
e-mail: guillermo.navarro@uab.cat

© Springer International Publishing AG, part of Springer Nature 2019
A. Said and V. Torra (eds.), *Data Science in Practice*, Studies in Big Data 46,
https://doi.org/10.1007/978-3-319-97556-6_7

security. In fact, privacy is often considered as a bargaining chip for security, and data privacy is claimed to be opposed to personalisation.

While it is true that more data can be useful for increasing security and personalization levels, it is also true that privacy permits us to enforce some security policies (as e.g., privacy in communication), and some studies show that the application of privacy mechanisms to data do not always degrade the quality of personalized services [29]. In addition to this, privacy friendly services can stand for an economic advantage.

The research on data privacy studies methods and procedures that avoid the disclosure of sensitive information. A large number of approaches exist based on different assumptions on the type of process that needs to be protected, the type of privacy model, and so on. Methods and measures also exist to evaluate data utility and to quantify the trade-off between privacy and utility.

In this chapter we present a short overview of the area. We focus on the aspects related to data science. For additional information see e.g., [36]. Some additional references are given in the conclusions section.

The structure of the chapter is as follows. In Sect. 7.2 we review privacy models and measures of disclosure risk. In Sect. 7.3 we give a summary of data protection mechanisms. In Sect. 7.4 we focus on differential privacy. In Sect. 7.5 we review masking methods and information loss measures to evaluate in what extent a masking method decreases data utility. The paper finishes with a summary and conclusions.

7.2 Privacy Models and Disclosure Risk Assessment

The literature discusses different types of disclosure that may occur when we release or publish a database. We can distinguish among two families of disclosure.

- Identity disclosure. When we release or publish a database we have identity disclosure when an intruder is able to locate the record of a certain individual in the database.
- Attribute disclosure. When we release some information, the intruder is able to learn something about some property attributed to some individuals.

It is important to underline that having identity disclosure does not mean that the intruder gets additional information on the individual identified in the database. In other words, identity disclosure does not always imply attribute disclosure. Consider the case that the intruder has more information on a person than what is available in the database. Nevertheless, it is expected that identity disclosure leads to attribute disclosure.

There are a number of different privacy models available in the literature. We discuss some of them below.

- Secure multiparty computation. We consider a set of parties that want to compute a joint function of their databases. The only information they want to share is the output of the function, but they neither want to share the data nor partial results. Cryptographic protocols are often used to implement this privacy model.

- Privacy from reidentification. This privacy model focuses on identity disclosure. When a database is released the goal is to avoid identity disclosure. Re-identification is the process of finding a record in the database. Given a file, masking methods modify this file in order to produce another file in which reidentification is not possible (or difficult).
- k-Anonymity [30, 31, 35]. This privacy model also focuses on identity disclosure. It is a special case of privacy from re-identification. A database satisfies k-anonymity when for each record in the database there are other $k - 1$ records that are indistinguishables. As in the case of re-identification, masking methods modify the original database to obtain one that satisfies the privacy model. For example, if we have a database X that includes information on age and town of some patients in a hospital, a k-anonymized version X' of this database X may use ages in ranges and counties instead of towns so that the change in granularity allow to have at least k patients for any combination of (age range, county).
- Differential privacy [12, 13]. This privacy model focuses on queries on a database. It wants to ensure that the output of a query does not depend (much) on the presence or absence of a record in the database. That is, $f(X)$ and $f(X')$ are similar when X and X' only differ in one record. Similarity is defined probabilistically. For example, if we want to compute the mean of X, f will be a randomized version of the mean so that for any record added or removed from X the mean does not change significantly.
- Integral privacy [37]. Let X and X' be a database and a version of this database after some editions (addition, suppression, and editing). Then, given functions $f(X)$ and $f(X')$ the goal is to avoid inferences on the databases and on the editions on these databases. Different assumptions are considered on the prior knowledge of the intruder. The definition is extended to the case of just X and $f(X)$.

Variations of these privacy models exist. Some add additional constraints (e.g., l-diversity adds attribute disclosure to k-anonymity) and others generalize these models (n-confusion). If we consider that k-anonymity is about building anonymity sets of cardinality k, n-confusion is a generalization in which the records of such anonymity sets are not required to be all equal.

Some of these approaches can be combined. For example, we can use a cryptographic protocol to compute a function satisfying differential privacy in a secure multiparty way. Similarly, several parties can decide to jointly compute a database compliant with k-anonymity or with reidentification privacy. Integral privacy can be combined with differential privacy (to define differintegral privacy).

7.2.1 Disclosure Risk Assessment

We have seen that there are different privacy models. Some of them define privacy as a Boolean condition. That is, the privacy model is either satisfied or not. Secure multiparty computation, k-anonymity and differential privacy follow this approach. Given k and a database, we can state if the database satisfies or not k-anonymity. Then, we can define data protection methods that given an arbitrary database build one that satisfies the privacy requirement. In this case, methods are compared with respect to the data utility. That is, given two methods that return a database compliant with the k-anonymity criteria, the best method is the one that modifies less the data and, thus, there is less information loss. Differential privacy follows a similar approach. It is based on a parameter ϵ, and then given ϵ we define a data protection method that answers the query in a way that the result is compliant with the differential privacy model. Again, the goal is to minimize the perturbation on the result.

Reidentification privacy does not follow this approach. It considers privacy as a measurable condition. Given a (protected) database, we can define a measure of risk in relation to the proportion of reidentifications that an intruder can obtain from the database. That is, if the intruder can reidentify half of the records, then we have a risk of 50%. In this case, data protection mechanisms (masking methods) are required to minimize the perturbation in a way that the risk is low. This results into a multicriteria optimization problem. Uniqueness is another way of assessing disclosure risk. In this case, risk is defined as the probability that rare combinations of attribute values in the protected data set are also rare in the original data set. Uniqueness was defined considering sampling as the data protection mechanism. Measures for reidentification and uniqueness are for assessing identity disclosure. There are also measures for assessing attribute disclosure. Rank-based and standard deviation-based interval disclosure are examples of such measures.

7.3 Data Protection Mechanisms

A large number of data protection mechanisms have been defined up to now. There are different ways to classify these methods.

- *On whose privacy is being sought.* In the usual scenario in data privacy we can distinguish three roles. The data respondents (the individuals that supply voluntarily or not their data), the data holder (or data controller, that accumulates and keeps the data from several individuals), and the data user. Here we understand the data users as the individuals that supply the data, but are concerned on their own privacy and, thus, proactively look for privacy solutions. For example, the user sends an email and in order to avoid any one accessing its content, encrypts it.

Whose privacy	Computation known?	Num. Sources	Protection methods
Respondent and holder	data-driven	multiple/one	Masking methods
Respondent and holder	computation-driven	multiple	Multiparty computation
Respondent and holder	computation-driven	one	Differential privacy
Holder	result-driven	one	Rule hidding
	Application	What to protect?	
User privacy	communication	identity	Mixes, crowds
User privacy	communication	data	Cryptography
User privacy	information retrieval	identity	P2P communities (Anonymous database search)
User privacy	information retrieval	data	DisPA, TrackMeNot, GooPIR

Fig. 7.1 Classification of data protection mechanisms according to the three dimensions: (i) whose privacy is being sought, (ii) what computation is being done, and (iii) number of data sources. For user privacy, classification is done according to (1) type of application, and (2) what we want to protect

Based on these three roles, we can consider respondent privacy, holder privacy, and user privacy. Note that both respondent and holder privacy are to be implemented by the data holder, but their interests are typically different. A client does not want others to know that she has bought 25 L of liquor in a week. While a supermarket director (data holder) may be interested for legal reasons to avoid disclosure of respondent's data (the shopping cart of this particular client), the real interest of the director is that competitors cannot use supermarket's data to increase their market share. Market shares do not depend on a particular client.

- *On the computations to be done.* The application of a data protection procedure can be tailored by the type of analysis to be applied to the data. Nevertheless, the analysis is not always known. We distinguish three cases. Computation-driven or specific purpose protection methods, which corresponds to the case that the function to be computed is known with detail. Data-driven or general purpose protection procedure, which corresponds to the case that the function is not known. In addition, we distinguish also result-driven protection methods which correspond to the case that we know the function (a data mining method) to be applied and what we want to protect is part of the results of this function. E.g., the supermarket wants to avoid that when researchers apply a rule mining algorithm to their database they mine a particular rule that give them a high competitive advantage.

- *On the number of data sources.* Some data protection procedures are defined for a single data source (a database) while others are about computing a function from several data sources (two or more databases). By taking time into account, the same database can give rise to several data sources, as is the case of multiple releases of the same database.

Existing data protection mechanisms can be classified according to these procedures. We describe them below. Figure 7.1 summarizes this classification.

Masking methods are the usual approach for data-driven methods for respondent and holder privacy. They correspond to methods to protect a database by means of reducing the quality of the database. That is, instead of the original database X a database X' is published with less quality. The masked database X' is published and, for example, researchers will be able to apply any machine learning algorithm to this data X'. This quality reduction is done by means of perturbative methods (that is, methods that change values introducing some kind of error) or non-perturbative methods (that is, methods that change the granularity of the data). Perturbative methods include noise addition and multiplication, rank swapping, microaggregation, and PRAM. Non-perturbative methods include generalization and suppression. In this latter case, the typical privacy model is reidentification or k-anonymity.

Methods for secure multiparty computation are used for computation-driven methods for respondent and holder privacy when multiple sources are considered. In this case, first we need to know in detail what function needs to be computed, and then we write a cryptographic protocol to compute this function ensuring that none of the parties (none of the data holders) learn anything but the final function. For example, a few supermarket directors want to compute which type of chocolate is the most popular. Nevertheless, they don't want to put the data of their sales together. In this case, the secure multiparty computation model is an adequate privacy model.

Methods to achieve differential privacy are used for computation-driven methods for respondent and holder privacy when a single source is considered. Differential privacy is typically either implemented by means of a perturbation of the original database and then computing the function from this perturbed data, or directly applying a perturbation to the output of the function. In this case, the privacy model is differential privacy.

User privacy methods are methods that empower the users of a system that is a data holder to protect the data that is produced from their actions within the system. For example, the users of a network can go together and form a mix network which will empower them with privacy over their connection history. Note that the communication history of a network is often stored as a database, therefore anonymous communication is a means for the users of a network to gain power over their privacy. User privacy can protect from identity disclosure or attribute disclosure. Methods for anonymous communication like mix networks [5] and Crowds [28] typically protect the identity of a user in a communication network, that is, identity disclosure. Attribute disclosure in anonymous communication can happen as in any other database. An example of an attribute is for example the connection time. Secrecy of the transmitted data could perhaps be considered protection from attribute disclosure, by considering the transmitted data to be the sensitive attribute. In any case, secrecy in communication is typically achieved by the use of cryptography. Another interesting application of user privacy is in information retrieval. For example, queries posted by P2P communities as in [33, 34] permit to protect identities in information retrieval. Agents that add noise to queries (as TrackMeNot [15] and GooPIR [10]) and systems that dissociate an identity into different virtual identities (as DisPA [17]) permit to protect the data (the queries) in information retrieval.

7.4 Differential Privacy

Differential privacy is to avoid inferences on the participation of an individual when computing a query. Then, the goal is that the output of this query is independent of the participation of an individual. Formally, it is defined in terms of the output of this query, which is supposed to be modeled in terms of a probability distribution. Then, given two databases that differ in one record, the corresponding probability distributions should be similar enough. The formal definition follows.

Definition 7.1 A function K_q for a query q gives ϵ-differential privacy if for all data sets D_1 and D_2 differing in at most one element, and all $S \subseteq Range(K_q)$,

$$\frac{Pr[K_q(D_1) \in S]}{Pr[K_q(D_2) \in S]} \leq e^\epsilon.$$

In this definition, ϵ corresponds to the level of privacy required. The smaller the ϵ, the greater the privacy we have. The selection of an appropriate parameter ϵ is an open problem, although some suggestions have been given in the literature. See e.g. [20] were the value of ϵ is discussed in relation to a reidentification context.

Differential privacy for numerical queries is often implemented in terms of adding noise to the actual output of the query. More specifically, the Laplace mechanism defines K_q in terms of the correct answer of query q and a random variable drawn from the Laplace distribution. The noise is based on the global sensitivity of the query.

Definition 7.2 [12] Let $\mathcal{D}$ denote the space of all databases; let $q : \mathcal{D} \to \mathbb{R}^d$ be a query; then, the sensitivity of q is defined

$$\Delta_{\mathcal{D}}(q) = \max_{D, D' \in \mathcal{D}} ||q(D) - q(D')||_1.$$

where $|| \cdot ||_1$ is the L_1 norm, that is, $||(a_1, \ldots, a_d)||_1 = \sum_i |a_i|$.

Then, $K_q(D) = q(D) + X$ where X is a random variable that follows a Laplace distribution of the form $L(0, \delta(q)/\epsilon)$.

This approach to differential privacy (i.e., that we have all the data, we compute q and then add some noise to correct answer to q) is also known as the centralized approach to differential privacy. In this case the data holder is the only one with access to the whole database, who computes $q(D)$ and introduces some noise into the output. Note that the data user does not have access to the database, only to the protected answer to a query.

An alternative to the centralized approach is local differential privacy. In this case, the data collector has data that has already been anonymized, and then computes the output from this anonymized data set. Or, equivalently, the data collector applies first an anonymization method to the original data set, deletes the original data set, and then computes the output from the anonymized data set. Local differential privacy for categorical data is often implemented in terms of randomized response. Given a binary attribute a randomized response is defined in terms of a probability of not modifying a and the probability of changing the value of a. Then, given any response for a, we will mask the value of a according to these probabilities. As randomized response change values, we have plausible deniability. Implementation of local differential privacy by means of randomized response is equivalent to the masking method PRAM discussed in the next section (see e.g. [39]). In general, local differential privacy for a query q is implemented as $K_q(D) = q(M(D))$ where $M(D)$ correspond to the application of a masking method.

7.5 Masking Methods

Given a database X, masking methods build another database X' which is similar to X in the sense that inferences from X and X' are similar. In addition, X' should avoid disclosure risk. Disclosure risk can be defined according to the privacy models discussed in Sect. 7.2. More particularly, masking methods are used with privacy from reidentification, k-anonymity, and differential privacy. Note that in this case, the data user has access to a protected database.

Masking methods are often classified into three main classes. They are the following ones.

- Perturbative methods. The database X' is defined as X and some noise that modifies the original values in X. That is, we can see X' as following the next equation: $X' = X + \epsilon$. Different perturbative methods exist. Noise addition is the simplest one [2]. In this case ϵ follows a normal distribution with zero mean and appropriate covariance. Noise multiplication [18], microaggregation [6, 9], rank swapping [23], post-randomization method (PRAM) [19], transformation-based methods are other examples of perturbative methods. The rationale for perturbative methods is that introducing some error the probability that an intruder finds a record decreases.

  Microaggregation is about finding sets of records that are similar and then replace their value by their average. The minimum number of records that need to contribute to the average is our privacy guarantee. PRAM is a method for categorical data and is defined in terms of a Markov matrix. Then, each category in the data set is replaced by another one (or not replaced) according to the probabilities in the Markov matrix. The Markov matrix is our privacy guarantee. As explained in Sect. 7.4, PRAM and randomized response can be seen as equivalent.
- Non-perturbative methods. In this case, given a database X, the new database X' is obtained by means of changes of granularity of the elements in X. For example, a number is replaced by an interval, or a term is replaced by a more general one (for example, a town is replaced by a county or a province, and the term widow is replaced by widow-or-divorced). Generalization (as in the previous examples)

and suppression (i.e., replace a value by an indication that the value has been suppressed) are two approaches of non-perturbative methods [35]. The rationale for non-perturbative methods is the same as for perturbative ones. In this case, one expects that reducing granularity reidentification is more difficult.

- Synthetic data generators. This approach consists of replacing X by data that has been generated artificially [4, 14, 27]. Partially synthetic and fully synthetic approches exist. The rationale of this approach is that synthetic data is not real data and, thus, reidentification cannot take place. Nevertheless, when the fit between the synthetic and the original data is too tight, then there may be disclosure risk anyway. This is certainly the case on partially synthetic data when only some attributes are synthetic, and also when each attribute is made synthetic by means of a model with respect to the others. In both cases, we will have one synthetic record for each original record, and both original and synthetic records may be similar.

7.5.1   Information Loss Measures

Given a database X and a protected version X' of this database, we need to know if X' is still useful for analysis. A few measures have been proposed to evaluate in what extent information has been lost in the masking process.

Information loss measures are defined in terms of a function of X and X' and the divergence between the values of this function. Formally, an information loss (IL) measure for function f and databases X and X' has the form

$$IL_f(X, X') = divergence(f(X), f(X')).$$

Different measures information loss measures have been defined in the literature, which differ in the type of function computed. Some try to give an evaluation of the divergence independent of the data use. They were based on some statistics for X as e.g. mean, variance, covariance, and correlations in the case of numerical data, and based on contingency tables for subsets of attributes in X or based on probabilities and entropies computed from X. Other measures have been defined with a particular data use. For example, there are measures that evaluate in what extent clusters in X and X' differ (using e.g. Rand and Jaccard indices) and measures that evaluate in what extent classifiers built from X and X' have the similar accuracy. For example, we can define $IL_{mean}(X, X') = \sqrt{(1/n) \sum_i (mean_i(X) - mean_i(X'))^2}$ if $mean_i$ corresponds to the mean of the ith attribute and there are n attributes in X.

In [7, 8] the authors propose a general use of statistical divergences between X and X. For example, the mean variation of the aggregation of the distance between records from X and X', their mean, and covariance with numeric data. This is very common for data-driven or general purposed masked data (cf. Sect. 7.3). Other similar generic information loss measure can be the use of entropy-based measures for discrete data [40] or other common statistical approaches (see discussion in [41]).

Similar approaches have also been translated to other types of data such as graphs or networks [3].

Depending on the type and purpose of the data to be masked, data specific information loss measures can be used. As previously mentioned, a common approach is to compare directly the outcome of the expected analysis to be performed in the data. Some examples are web mining [22], user profiling from query logs [25], graph mining [24, 26], or natural language processing [1, 32].

We note that although there are some generic information loss measures, there are no widely used information loss models such as there are for disclosure risk (see Sect. 7.2). Information loss is very dependent on the type of data to be protected and the purpose of such data.

7.6 Summary

In this chapter we have given an overview of different aspects related to data privacy. We have discussed privacy models, ways to evaluate disclosure risk and information loss measures.

From a data science point of view, respondent and holder privacy is more relevant than user privacy. Methods to deal with the former type of privacy will be the ones that data scientists may need to take into account and implement. Because of that, masking methods, methods for secure multiparty computation and methods to achieve differential privacy may be needed in applications.

Additional information on data privacy can be found in e.g. [11, 16, 21, 36, 38]. The first reference [36] gives a general perspective of the whole area of data privacy. Then, [11, 16] focus on privacy for statistical databases. In contrast, [38] is on privacy for data mining with a focus on the secure multiparty computation model. Finally, a suitable reference for further reading on differential privacy is [21].

References

1. Anandan, B., Clifton, C., Jiang, W., Murugesan, M., Pastrana-Camacho, P., & Si, L. (2012). t-Plausibility: Generalizing words to desensitize text. *Transactions on Data Privacy, 5*(3), 505–534.
2. Brand, R. (2002). Microdata protection through noise addition. In *Inference control in statistical databases* (pp. 97–116). Springer.
3. Casas-Roma, J., Herrera-joancomartí, J., & Torra, V. (2013). Analyzing the impact of edge modifications on networks. In: *The 10th International Conference on Modeling Decisions for Artificial Intelligence* (Vol. 8234, pp. 296–307). Lecture notes in computer science. Springer.
4. Cano, I., & Torra, V. (2009). Generation of synthetic data by means of fuzzy c-regression. In *Proceedings of IEEE International Conference on Fuzzy Systems* (pp. 1145–1150).
5. Chaum, D. L. (1981). Untraceable electronic mail, return addresses, and digital pseudonyms. *Communications of the ACM, 24*(2), 5.

6. Defays, D., & Nanopoulos, P. (1993). Panels of enterprises and confidentiality: The small aggregates method. In *Proceedings of the 1992 Symposium on Design and Analysis of Longitudinal Surveys, Ottawa: Statistics Canada* (pp. 195–204).

7. Domingo-Ferrer, J., & Torra, V. (2001). A quantitative comparison of disclosure control methods for microdata. In *Confidentiality, disclosure and data access: Theory and practical applications for statistical agencies* (pp. 111–134).

8. Domingo-Ferrer, J., Mateo-Sanz, J. M., & Torra, V. (2001). Comparing SDC methods for microdata on the basis of information loss and disclosure risk. In *Pre-proceedings of ETK-NTTS, 2001* (Vol. 2, pp. 807–826).

9. Domingo-Ferrer, J., & Mateo-Sanz, J. M. (2002). Practical data-oriented microaggregation for statistical disclosure control. *IEEE Transactions on Knowledge and Data Engineering, 14*(1), 189–201.

10. Domingo Ferrer, J., Solanas, A., & Castellà Roca, J. (2009). h(k) private information retrieval from privacy uncooperative queryable databases. *Online Information Review, 33*(4), 720–744.

11. Duncan, G. T., Elliot, M., & Salazar, J. J. (2011). *Statistical confidentiality.* Springer.

12. Dwork, C. (2006). Differential privacy. In *Proceedings of ICALP 2006* (Vol. 4052, pp. 1–12). LNCS.

13. Dwork, C. (2008). Differential privacy: A survey of results. In *Proceedings of TAMC 2008* (Vol. 4978, pp. 1–19). LNCS.

14. Fienberg, S. E., Makov, U. E., & Steele, R. J. (1998). Disclosure limitation using perturbation and related methods for categorical data. *Journal of Official Statistics, 14*(4), 485–502.

15. Howe, D., & Nissenbaum, H. (2009). TrackMeNot: Resisting surveillance in web search. In *Lessons from the identity trail: Anonymity, privacy, and identity in a networked society.* Oxford University Press.

16. Hundepool, A., Domingo-Ferrer, J., Franconi, L., Giessing, S., Nordholt, E. S., Spicer, K., & de Wolf, P. -P. (2012). *Statistical disclosure control.* Wiley.

17. Juàrez, M., & Torra, V. (2015). DisPA: An intelligent agent for private web search In G. Navarro-Arribas, V. Torra (Eds.), *Advanced research on data privacy* (pp. 389–405). Springer.

18. Kim, J. J., & Winkler, W. E. (2003). Multiplicative noise for masking continuous data (Research Report Series No. Statistics #2003-01). Statistical Research Division. U.S. Bureau of the Census.

19. Kooiman, P., Willenborg, L., & Gouweleeuw, J. (1998). *PRAM: A method for disclosure limitation of microdata.* Research Report, Voorburg: Statistics Netherlands.

20. Lee, J., & Clifton, C. (2011). How much is enough? Choosing ϵ for differential privacy. In *Proceeding of ISC 2011* (Vol. 7001, pp. 325–340). LNCS

21. Li, N., Lyu, M., Su, D., & Yang, W. (2016). *Differential privacy: From theory to practice.* Morgan and Claypool Publishers.

22. Navarro Arribas, G., & Torra, V. (2010). Privacy preserving data mining through Microaggregation for Webbased E-commerce. *Internet Research, 20*(3), 366–84.

23. Moore, R., (1996). *Controlled data swapping techniques for masking public use microdata sets.* U. S. Bureau of the Census (unpublished manuscript).

24. Mülle, Y., Clifton, C., & Böhm, K. (2015). Privacy-integrated graph clustering through differential privacy. In *EDBT/ICDT Workshops* (pp. 247–254).

25. Navarro-Arribas, G., Torra, V., Erola, A., & Castellà-Roca, J. (2012). User K-Anonymity for privacy preserving data mining of query logs. *Information Processing & Management, 48*(3): 476–487. (May 2012).

26. Nettleton, D. F. (2012). Information loss evaluation based on fuzzy and crisp clustering of graph statistics. *IEEE International Conference on Fuzzy Systems* (pp. 1–8).

27. Raghunathan, T. J., Reiter, J. P., & Rubin, D. (2003). Multiple imputation for statistical disclosure limitation. *Journal of Official Statistics, 19*(1), 1–16.

28. Reiter, M. K., & Rubin, A. D. (1998). Crowds: Anonymity for web transactions. *ACM Transactions on Information and System Security, 1*(1), 66–92.

29. Sakuma, J., & Osame, T. (2018). Recommendation with k-Anonymized Ratings. *Transactions on Data Privacy, 11*(1), 47–60.

30. Samarati, P., & Sweeney, L. (1998). *Protecting privacy when disclosing information: k-anonymity and its enforcement through generalization and suppression*. Rep: SRI Intl. Tech.
31. Samarati, P. (2001). Protecting respondents' identities in microdata release. *IEEE Transactions on Knowledge and Data Engineering, 13*(6), 1010–1027.
32. Sánchez, D., & Batet, M. (2017). Toward sensitive document release with privacy guarantees. *Engineering Applications of Artificial Intelligence, 59*(Supplement C), 23–34.
33. Stokes, K., & Bras-Amorós, M. (2011). On query self-submission in peer-to-peer user-private information retrieval. In *Proceedings of 4th PAIS 2011*.
34. Stokes, K., & Farràs, O. (2014). Linear spaces and transversal designs: k-anonymous combinatorial configurations for anonymous database search. *Designs, Codes and Cryptography, 71*, 503–524.
35. Sweeney, L. (2002). Achieving k-anonymity privacy protection using generalization and suppression. *International Journal of Uncertainty, Fuzziness and Knowledge-Based Systems, 10*(5), 571–588.
36. Torra, V. (2017). *Data privacy*. Springer.
37. Torra, V., & Navarro-Arribas, G. (2016). Integral privacy. In *Proceedings of CANS 2016* (Vol. 10052, pp. 661–669). LNCS.
38. Vaidya, J., Clifton, C. W., & Zhu, Y. M. (2006). *Privacy preserving data mining*. Springer.
39. Van den Hout, A. (2004). *Analyzing misclassified data: Randomized response and post randomization*. Ph.D. thesis, Utrecht University.
40. Willenborg, L., & de Waal, T. (2001). *Elements of statistical disclosure control*. Springer.
41. Winkler, W. E. (2004). Masking and re-identification methods for public-use microdata: Overview and research problems. In *Privacy in statistical databases* (pp. 231–246). Springer.

Chapter 8
Visual Data Analysis

Juhee Bae, Göran Falkman, Tove Helldin and Maria Riveiro

Abstract Data Science offers a set of powerful approaches for making new discoveries from large and complex data sets. It combines aspects of mathematics, statistics, machine learning, etc. to turn vast amounts of data into new insights and knowledge. However, the sole use of automatic data science techniques for large amounts of complex data limits the human user's possibilities in the discovery process, since the user is estranged from the process of data exploration. This chapter describes the importance of Information Visualization (InfoVis) and visual analytics (VA) within data science and how interactive visualization can be used to support analysis and decision-making, empowering and complementing data science methods. Moreover, we review perceptual and cognitive aspects, together with design and evaluation methodologies for InfoVis and VA.

8.1 Introduction—Why Visualization?

The access to ever increasing amounts of data comes with the promise of more accurate and effective decisions. However, the dream of automating the analysis processes is hindered by factors such as noisy and uncertain data, together with the fact that many problems are ill-defined, making fully automatic solutions unsuitable or even infeasible. Information visualization enables human analysts to analyze data without knowing exactly which questions to pose in advance, enabling exploratory analyses to take place. Through visualization, the analyst is aided in the process of

J. Bae · G. Falkman · T. Helldin (✉) · M. Riveiro
Skövde Artificial Intelligence Lab, School of Informatics, University of Skövde,
Skövde, Sweden
e-mail: tove.helldin@his.se

J. Bae
e-mail: juhee.bae@his.se

G. Falkman
e-mail: goran.falkman@his.se

M. Riveiro
e-mail: maria.riveiro@his.se

© Springer International Publishing AG, part of Springer Nature 2019
A. Said and V. Torra (eds.), *Data Science in Practice*, Studies in Big Data 46,
https://doi.org/10.1007/978-3-319-97556-6_8

detecting patterns and trends in the data, something we as humans are very good at through our visual apparatus, leaving the complex computations necessary to be performed by the computer, thus combining the strengths of humans and computers.

Visualizations can be used at different stages in the analysis process. In a first step, it can be used to better understand the data dealt with, aiding in the process of choosing analysis strategies and methods. Visualizations can also be used to refine the analysis method used by exploring the effects of, for example, parameter changes. Of course, visualizations are often used to aid end-users make decisions based on the data, where factors such as the validity of the analysis results play an important role.

In this chapter, we first outline important aspects of the human visual system, enabling the effective usage of information visualization techniques. Moreover, we present the research areas of Information Visualization (InfoVis) and visual analytics (VA), outlining important aspects of how data scientists can use visualizations to incorporate human expert knowledge in the analysis process. The chapter is concluded with guidance how you as a data scientist can evaluate your visual solutions.

8.2 Perception and Cognitive Aspects of Viusalization

"A picture is worth ten thousand words" the famous proverb says, and visualizations, or graphic representations, can indeed be seen as information highways that enable very fast transformation of what we visually perceive into information, knowledge and insight. Before we continue, we need a basic understanding of the infrastructure of these highways: how we constantly pose *visual queries* in order to solve tasks using visual perception, how our *visual apparatus* supports this, how answers are found through a *visual search process*, and the important roles our *channels of attention* and our *perception of color* play in this.

8.2.1 Visual Queries

We do not build and maintain a coherent and comprehensive mental picture of our environment. Instead, we constantly sample the visual space surrounding us on a "need-to-know" basis. This means that we have mechanisms for quickly getting access to visual information that might be of interest and comparing this to what we are looking for as given by the goals or tasks at hand. The first part has much to do with *attention* and our ability to operate our body to acquire the information that interests us. The second part has much to do with our ability to quickly test and *recognize visual patterns*.

In [70], "visual thinking" was described as a series of *visual queries* that guide our visual apparatus. As an example, using a road map, our task is to find the quickest way from Skövde to Stockholm. Unconsciously, our eyes are drawn to certain parts of the map that immediately captures our attention, such as larger dots (cities) and lines (roads). At the same time, our task makes us start forming visual queries, such as "where are the heaviest lines?" (major highways) and "where is the text pattern 'Stockholm'?" (the destination). This way, the queries prime our visual apparatus to focus on finding the desired patterns. If we find a match, part of the solution has been found, and we move on forming a new query. If not, we move our eyes (and head) to look at other parts of the map. Eventually, all queries have been answered and we have found a way from Skövde to Stockholm.

8.2.2 The Visual Apparatus

Starting with what would correspond to the image sensor of a digital camera, the eye consists of roughly 125 million photoreceptor cells. There are two types of photoreceptors: rods and cones. Rods are more sensitive to luminance and motion but not at all to color, they have lower spatial acuity but have shorter response time, and they are much more numerous, especially at the perimeter of the retina. Cones have lower light sensitivity but are chromatic and have higher spatial resolution, they constitute only about 5% of the total number or receptors but are the only ones to be found in the very focus of the eye—the *fovæ centralis*. As a consequence, to perceive the color of objects at a high level of detail the objects must be in the center of the visual field. To best perceive flickering or moving objects with low contrast relative to the background the objects should be near the edge of the visual field.

The different amounts of light and color data captured by the eye do not, however, directly correspond to the sizes of the areas of the brain cortex where this data is processed. There is a step of data compression, or data fusion, when the signals from ∼125 million receptors are mapped to the approximately 1 million fibres of the optic nerve. The degree of compression is much lower in the fovæ area (∼5 cells per fibre) compared to the perimeter of the retina (thousands of cells per fibre). This means that more than 50% of the brain's processing power is devoted to less than 5% of the captured optic data. As a result, at arm's length, we only see sharply in an area corresponding to the size of a thumb's nail, but at the very center of that area we can perceive 100 dots on the head of a needle. In contrast, at the edge of our peripheral vision we can only perceive objects the size of a fist at arm's length.

Visual thinking as a series of visual queries means that we do not look at our environment in fixed steadiness; instead, the eyes constantly move around, looking for regions and details of interest. These jumps from one eye position to another— referred to as a *saccade*—are extremely fast, reaching angular speeds up to 900°/s. A saccade normally takes ∼200 ms to initiate, and then takes about 20–200 ms to execute, depending on the amplitude, but, as we will see, initiation time can be greatly reduced by priming the visual apparatus. This is important since during a saccade we are subject to so called *saccadic masking*, rendering us effectively blind (effects that would result in blurriness or gaps in visual perception are suppressed).

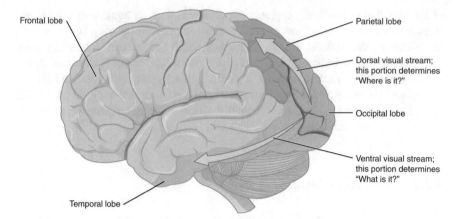

Frontal lobe

Parietal lobe

Dorsal visual stream; this portion determines "Where is it?"

Occipital lobe

Ventral visual stream; this portion determines "What is it?"

Temporal lobe

Fig. 8.1 The visual pathways for "what" and "where" queries. (Anatomy and Physiology, Connexions Web site. http://cnx.org/content/col11496/1.6/, Jun 19, 2013, by OpenStax College Data is licensed under CC Attribution 3.0 Unported.)

The signals from the optic nerve ends up at the back of the brain, in the *primary visual cortex*, also referred to as *visual area 1*, V1, from which information is further relayed to subsequent visual areas, V2–V6. V1 and V2 are concerned with the detection of basic features of the visual space, such as size, edges, color, motion, orientation, form and depth. The further up in the chain of visual areas, the more complex visual representations, or patterns, are formed. For instance, the *lateral occipital complex* (LOC) is where general shape detection occurs, which the *inferior temporal cortex* (IT) subsequently uses for recognizing objects, such as faces, number and letters. There are two general pathways of visual processing, and which is used depends on the visual query in question: "What" queries, i.e., queries associated with form recognition, object representation and long-term memory, follow the *ventral* pathway; "where" and "how" queries, i.e., queries associated with object location and motion, or requiring controlling the body in the answering of the query (especially movement of the eyes and the head), follow the *dorsal* pathway. See Fig. 8.1.

For a more detailed account on the visual apparatus and its influence on visual perception we refer to [70].

8.2.3 The Visual Search Process

The visual processing pathways described above are actually two parallel processes: In a bottom-up process, visual information in the form of light patterns on the retina drives pattern building from basic features. In a top-down process, the visual queries at hand and their corresponding need for attention in terms of objects and patterns of interest reinforce perceptual actions (e.g., eye movements) and prime (tune) signals in the bottom-up chain.

At a conceptual level, the *visual search process* can be described in terms of three nested loops [69]:

Pattern testing In the innermost loop, information from the current area of *fixation*—the period between saccations—is tested against patterns of interest, at a speed of approximately 20 patterns per second. Given that a fixation normally lasts for about 0.2 s, this means that ~4 patterns are tested in each iteration.

Search for patterns If we fail to find the pattern of interest in the current fixation area, the second loop starts by priming the bottom-up process towards relevant features. It also reinforces eye movements by remembering fixation points already visited in the area in order to avoid repeated testing. As a final step, a new nearby fixation candidate is identified in the area and eye movements are made to acquire it.

Identify candidates If the second loop fails to find a fixation candidate, the search for visual elements (patterns, objects etc.) of relevance to the visual queries at hand must be extended. The outermost loop starts by priming the top-down process towards relevant patterns and objects, using task and domain knowledge and experience from similar visual queries. It also reinforces actions needed to perceive new, more peripheral, parts of the environment in order to identify candidate areas. Finally, eye, head and possibly also body movements are made to acquire a new area of fixation.

8.2.4   Channels of Attention

An important part of the visual search process is the constant tuning of visual signals in order to strengthen our attention to objects and patterns of interest. What guides and influences the processing in visual areas V1 and V2 is coupled to what features "stand out" or "distract"—what draws our perceptual attention—in the visual space. Research (cf. [64]) has shown that the following *pre-attentive features* are tunable in V1 and V2, in the meaning that they can be used to reinforce the forthcoming eye movements and the search for and testing of patterns:

- Color (hue and lightness).
- Elementary shape (size; elongation; curvature; sharpness/fuzziness).
- Orientation (direction; angulation; alignment/misalignment).
- Motion (direction and phase of motion; blinking).
- Spatial grouping (proximity; joined lines; enclosing contours or color region).
- Depth (shadow; convex/concave).

How well an object or pattern stands out—the *pop-out effect* it generates—does not depend on the number of distractive features it exhibits. It is the relative contrast to other objects or patterns that is decisive for our attention being drawn to the object or not, and the relative contrast must be above a certain threshold value for the pop-out

Fig. 8.2 Non-symmetrical pop-out effect: Left image: The area of the central dot is half the area of the rest of the dots. Right image: The area of the central dot is double the area of the rest of the dots. The resulting pop-out effect is clearly larger in the right image as compared to the left image

effect to occur. Also, the generated pop-out effect is often not symmetric: an increase in, e.g., shape size, will make an object stand out more clearly from other objects than an equally large decrease in shape size would have made (see Fig. 8.2).

The above features are "hard-coded" in V1 and V2, both in the sense of our attention automatically and unconsciously being drawn to patterns and objects that stand out in terms of these features (we cannot train ourselves not to immediately see the single red dot in a large collection of otherwise black dots, or directly notice the single outlier of an otherwise dense cluster of objects), and in the sense that these features are processed separately in V1 and V2, thus forming separate *channels of attention*. In every moment, V1 and V2 can be tuned to a pre-attentive feature, making it easy to, e.g., find all squares in a collection of circles and squares by priming the elementary shape channel or to find all green objects in a collection of red and green objects by priming the color channel. Although some conjuncts of features are tunable, most conjuncts are not pre-attentive (does not generate a pop-out effect). This explains why it is much harder to find all green squares in a collection of red or green circles and squares, since that would require priming two different channels of attention at the same time, and *shape ∩ color* is not a pre-attentive conjunct.

By encoding objects and patterns using visual features from different channels we can simultaneously pay attention to several objects and patterns: one type of object could be encoded using color, another using shape, a third using motion and so on. To this we can add the variants within each channel: one type of object can be encoded using hue, another using lightness and so on. Experience show that a handful of channel features can be combined this way, but it is very hard to effectively use more than 8–10 features within a single visualization or graphic representation.

8.2.5 Perception of Color

As described in Sect. 8.2.2, cones are sensible to color: some cones are sensitive to mostly long wavelength light (reds), some to mostly medium wavelength light (greens), while other cones are sensible to mostly short wavelength light (blues). The information from the cones (and the rods) are processed in V1, where the signals are added and subtracted in several steps, thereby forming three *color-opponent channels*: red–green, blue–yellow and black–white (the last one being achromatic, only detecting luminance). A strong signal in one direction on one channel and neutral signals on the other two channels result in the six *psychological primary colors* black, white, red, green, yellow and blue, so called because any other color could be described in terms of some combination of these. Research within linguistics has shown that there seems to be a consensus among different cultures on these being "true" primary colors, and also that there is no ordering of these colors (i.e., black does not "come before" white, blue is not "larger than" green etc.).

Compared to the other two channels, the black–white channel has much higher capacity to encode detailed visual information (much higher resolution), especially regarding spatial information, but also when it comes to motion and depth. Thus, a high-density visualization or graphic representation should use grayscale to encode objects and patterns, as opposed to any other coloring scheme.

An important observation is that the interpretation of color is context sensitive, and is influenced both by nearby colors, nearby difference in luminance, nearby shadows and nearby textures. As a consequence, two objects that, according to the color-opponent channel model, have exactly the same color, but are situated in two different context, might be perceived to have two very different colors.

8.3   Information Visualization

Visual representations help people to understand abstract data. For instance, tracing the flight [33] uses live data to visualize thousands of commercial flights on a geographical map. If displayed in simple text and numerical information, it would be much more difficult to identify hub areas with high traffic volumes. According to many definitions of information visualization, it uses the computer support to enhance human's cognition. Two early definitions of Information Visualization (InfoVis) are:

- *The use of computer-supported interactive, visual representation of abstract data to amplify cognition* [13].
- *Information visualization utilizes computer graphics and interaction to assist humans in solving problems* [21].

This section gives an overview of the research area of information visualization, providing the reader with references for further exploration.

8.3.1 InfoVis Tools and Applications

A lot of tools are now available online with the help of practitioners and researchers providing open source code together with tutorials. Some of the examples are d3.js [11], javascript infovis toolkit [7], vega [53], prefuse [23], plotly [24], gephi [5], and raw [38]. And surely, we have been using histograms, pie charts, bar graphs, scatter plots, radial graphs, and treemaps [56] in many areas such as in industry and academia.

The areas that apply information visualization, to name some, are scientific research, digital libraries, data mining, information graphics, financial data analysis, marketing, manufacturing production control, crime mapping, etc. [6].

8.3.2 InfoVis as a Research Field

Information visualization is a method to visually communicate the information efficiently and effectively, both aesthetically and functionally. It can be used to explore the knowledge field, to support the decision-making process, to confirm a model and certain dataset, or to simply present results.

A renowned work in the early days is hold by Bertin [8, 9], when his theory obtained a significant position in the information design area. He describes the use of signs and symbols for two-dimensional static presentations based on practical experience, but unfortunately not thoroughly evaluated. The concept of information visualization has evolved from static presentations to more dynamic representations since then. Bederson and Shneiderman [6] remark that the field of information visualization has emerged "*from research in human-computer interaction, computer science, graphics, visual design, psychology, and business methods. It is increasingly applied as a critical component in scientific research, digital libraries, data mining, financial data analysis, market studies, manufacturing production control, and drug discovery*".

Nowadays, it is encouraged to communicate with more intuitive and even insightful visuals. Among many surveys within the field, Tufte [65] provides inspiring cases of various graphical examples. Few [19] introduces practical data visualizations for analysis and Ware [70] guides the ways to design visuals and the relations to human perception. Recently, Munzner [44] offers a synthesis view on the visualization field with extensive models and frameworks during the past 15 years. With the knowledge we gained over the decades, the beginners are able to learn from a collection of existing visuals where some of the tools are open-source and already available online.

We are used to techniques such as tables, histograms, pie charts, bar graphs (Fig. 8.4j), and scatter plots (Fig. 8.4b). However, there is a higher demand on more intuitive and insightful visuals using interactions. In fact, with the information flood and easier accessibility, we face even higher needs to communicate with

(a)   (b)   (c)

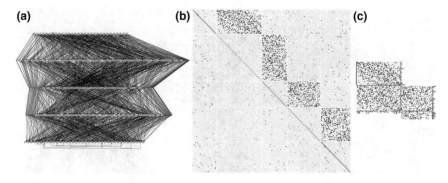

Fig. 8.3 Multiple visualization techniques displayed in the same dataset. **a** Node-link diagram, **b** matrix view, **c** quilts [4]

multi-dimensional and heterogeneous dataset. Some of the examples are time-series data (using i.e. index charts, stacked/stream graphs (Fig. 8.4l)), statistical distributions (using i.e. scatter plot matrices, parallel coordinates (Fig. 8.4i)), maps (using i.e. flow maps, choropleth maps), hierarchies (using i.e. node-link (Fig. 8.3a), adjacency diagrams) and networks (using i.e. force-directed layouts (Fig. 8.4e), matrix views (Fig. 8.3b)) [22].

Information visualization can be grouped based on data characteristics, techniques, and dimension management. A dataset can include only numbers, only text, graphs, or a combination of them. Ward et al. [68] provide pseudocode and algorithms of various types of visualization techniques with a bottom-up viewpoint on visualization. The authors divide techniques based on the data types: spatial, geospatial, multivariate, trees/graphs/networks, and text/document. They also demonstrate interaction concepts and related interaction techniques. In Liu et al. [35] work on visualization is categorized in four categories based on the data used: graph, text, map, and multivariate data visualizations. They further present a taxonomy of InfoVis techniques used during recent years.

In addition, users and viewers are able to visualize multiple visuals in one frame with the help of greater accessibility to open source codes and various techniques and tools. For instance, relationships of automobile features [22] on horsepower, weight, acceleration, and displacement can be represented in a scatter plot matrix and parallel coordinates. Also, a code package hierarchy [22] can be viewed in a tree diagram, a cartesian node-link diagram, a sunburst (radial space-filling), and a treemap. Figure 8.3 illustrates an example of a synthetic dataset displayed with a node-link diagram (Fig. 8.3a), a central matrix view (Fig. 8.3b), and quilts (Fig. 8.3c) with the same dataset [4]. Using the linking and brushing interaction techniques, i.e. where a change in one visualization is also reflected in the others, thus connecting multiple visualizations, can help to overcome the limitations of a single visualization technique [28].

To date, there are numerous venues for InfoVis research. The main resources within the field are:

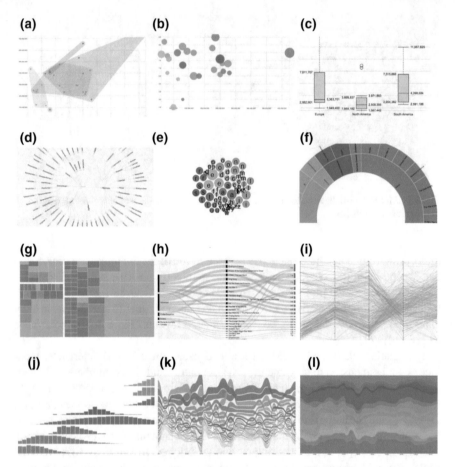

Fig. 8.4 Information visualization examples. **a** Convex hull, **b** scatter plot, **c** box plot, **d** circular dendrogram, **e** clustered force layout, **f** sunburst, **g** treemap, **h** alluvial diagram, **i** parallel coordinates, **j** bar chart, **k** bump chart, **l** streamgraph. Adapted from Mauri et al. [38]

- Journals: ACM SIGGRAPH, IEEE Transactions on Visualization and Computer Graphics, Computer Graphics Forum, IEEE Computer Graphics and Applications, Information Visualization.
- Conferences: IEEE Symposium on Information Visualization (InfoVis) (published as a special issue of IEEE TVCG since 2006), IEEE Pacific Visualization Symposium (PacificVis), EuroVis, International Conference on Information Visualization.

8.3.3   InfoVis Research Challenges

Below, we summarize InfoVis research challenges (which are also relevant to recent visual analytics, see Sect. 8.4). Since visualization and visual analytics are comprehensively connected, establishing and measuring evaluation methods and trustworthiness are challenges in both areas.

- *Structure new methodologies for evaluation metrics and obtain better understanding of users and tasks.*

  There has been criticism on using only time and error metrics to validate the performance of a visualization tool or system. Kosara [34] demonstrates memorability and engagement as potential new metrics. Yet, how can we measure these in an effective way and what are the tradeoffs? We find that there is an increasing number of publications performing data exploration and analysis as well as follow-up user studies and empirical evaluations [35]; however, there still remain unsolved issues [35]. Moreover, Chen [16] describes the need for more attention to better understand the elementary perceptual-cognitive tasks. In addition, it is not clear how the users' prior knowledge affects in an effective dialog between information visualization and its users.

- *Enhance trust and interpretability.*

  As viewers or users are more involved throughout the visual interaction and analysis process, incorporating visual trust [37, 39] is becoming more important to enhance user knowledge and cognition ability. However, there seems to be different levels and meanings of trust which can also be measured differently. Endert et al. [18] provide views on both qualitative and quantitative levels.

- *Perform interdisciplinary research with other areas, i.e. machine learning.*

  There is an increased trend to perform interdisciplinary research [18, 30]. For instance, final results are rendered in a static way in machine learning, but we see more research work to incorporate visualization and interaction mechanisms for a better understanding. Recently, there has been more communication among researchers in different areas in order to solve problems together (i.e. render intermediate results (TensorFlow [1], t-SNE [66]) and mechanisms to manipulate algorithm parameters).

- *Others*

  Other issues include integrating heterogenous data and displaying visual scalability and streaming data [35]. Visualizing causality [3] and uncertainty, measuring intrinsic quality, and gaining visual inference and prediction [16] are also issues to be solved.

8.4   Visual Analytics

Exploring and analyzing large amounts of data is becoming more and more difficult. The data is often disparate, dynamic and conflicting, as well as stemming from various heterogeneous sources. The sole use of automatic analysis methods reduces the human analysts possibilities to understand the data, as well as to input his/her expertise in the analysis. Further, automatic analyses only work well for well-defined and specific problems [32], making such approaches unsuitable for explorative analyses where analysts are searching for insights from data without predefined hypotheses. Moreover, today it is not only trained data scientists who are to analyze this data, but domain experts and decision makers with various backgrounds. Visual Analytics (VA) has been recognized as a way of enabling more effective understanding and analysis of large datasets, based on the assumption that interactive visual representations can augment the human capabilities of detecting patterns and making inferences [26]. By taking advantage of the analytical capabilities of the computer and the creativity of the human analyst, VA sets the foundation for enabling novel and unexpected discoveries.

This section provides a short overview of the research field of VA, outlining its importance within data science.

8.4.1   VA Definition and Process Model

VA strives to facilitate the analytical process by creating software that enables the human analyst to make use of his/her capability to perceive, understand and reason about the data. The VA process is characterized by the interaction between data, visualizations, data models, and the analysts to discover knowledge [27]. Figure 8.5 gives an overview of the stages (ovals) and their transitions (arrows) in the VA process. The first step is to pre-process and transform the data through data cleansing, grouping and normalization activities. After the data transformation stage, the analyst can choose to apply automatic or visual analysis methods. If an automatic approach is chosen, the analyst may choose among various data mining methods to analyze the input data and create a model of it. After the model creation, the analyst needs to evaluate and refine the model, often through interactive means where input data, parameters and/or analysis algorithms can be altered. Through model visualizations, the analyst can evaluate the findings of the generated models, ultimately leading to more and more knowledge about the data and problem at hand. If visual data exploration had been chosen first, the analyst has to confirm the generated hypotheses by an automated analysis. Misleading results at an early stage in the analysis can thus be detected, leading to better and more transparent results where the analyst can zoom in on different data areas or consider different visual views of the data. As such, compared with fully automatic analyses, the human analyst is in control of the analytical process.

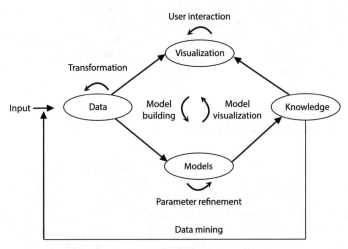

Visual Data-Exploration

Fig. 8.5 The visual analytics process (adapted from [27]). Data is collected, transformed and analyzed by automatic and manual means, incorporating the human analyst in the reasoning carried out

VA has grown from information visualization, but includes areas such as analytical reasoning, decision-making, data analysis and human factors. According to Thomas and Cook [62], VA involves the following areas:

- Analytical reasoning techniques that let users obtain deep insights which support assessment, planning and decision-making,
- Visual representations and interaction techniques that exploit the human eyes broad bandwidth pathway into the mind to let users see, explore and understand large amounts of information simultaneously,
- Data representations and transformation that convert all types of conflicting and dynamic data in ways that support visualization and analysis, and
- Techniques to support production, presentation and dissemination of analytical results to communicate to audiences.

As such, VA is highly interdisciplinary, focusing on the integration of all of these diverse areas. Due to its multi-disciplinary nature, defining VA is not easy. An early definition was "the science of analytical reasoning facilitated by interactive human-machine interfaces" [61]. However, based on its current practice, a more specific definition would be: "visual analytics combines automated analysis techniques with interactive visualizations for an effective understanding, reasoning and decision-making on the basis of very large and complex datasets" [26].

Central to VA is Shneiderman's [57] famous visual information seeking mantra "overview first, zoom/filter, details on demand". This mantra was extended by Keim et al. [29] to depict the VA process framework: "analyze first, show the important, zoom/filter, analyze further, details on demand", indicating that it is not sufficient to just collect and display the data visually, but rather that is necessary to analyze the data according to its value of interest, to show the most relevant aspects of the data, as well as to allow the user to get details of the data when needed through the provided interactions.

8.4.2   VA Tools and Applications

To support the VA process, several tools and applications have been developed which incorporate data mining and visualization functionalities. Examples of general, commercial tools are Tableau (www.tableau.com), Spotfire (www.spotfire.com), SAS (www.sas.com) and Datameer (www.datameer.com), whereas a myriad of application specific tools also can be found related to, for instance, biology [14], security [20, 36] and geography [58]. As a result, VA is gaining more and more attention from both industry and academia.

8.4.3   VA As a Research Field

The formal beginning of VA is often marked by the publication "Illuminating the Path: The R&D Agenda for Visual Analytics" [61] in 2005, where researchers within the field identified challenges and main VA science areas: the science of analytical reasoning, visual representations and interaction techniques, data representations and transformations and presentation, production and dissemination. Forums for VA publications are, amongst others, IEEE Visual Analytics Science and Technology (VAST) Symposium (www.ieeevis.org), information visualization journals, conferences and related workshops.

The initial domain driving the development of VA was homeland security where the focus was on supporting the detection of, response to and recovery from man-made and natural disasters. Consequently, some of the first deployments for VA technology have been for the public safety and emergency response communities. However, other fields rapidly gained interest in VA, especially in domains where increasing data volumes, complex data analysis and the need to communicate analytic outcomes were prevalent, such as human and environmental health, economics, and commerce [63]. In a recent survey of VA applications and research [59], it was concluded that the main efforts are conducted regarding the design of intuitive user interactions and the visual mappings of existing algorithms, than regarding the interaction with the models used in the analysis.

8.4.4 VA Research Challenges

Several researchers within the VA community have listed a range of different challenges that need to be addressed to guide the future development of VA applications (see for instance [26, 29, 59, 63]). A summary is provided below:

- Scalability and data management: a great challenge lies in the ability to visualize big data interactively. Many existing VA techniques and tools are able to handle small/medium sized data, but are not scalable to extreme-scale data. Many analytic applications use in-memory storage rather than a database approach, making the analysis a tedious process. Dealing with streaming data also presents challenges, and much time and effort is often needed to deal with low quality, missing, incomplete and/or erroneous data, which adds to the complexity of integrating data from many sources.
- Black-box models: to communicate the findings from complex models to enable sensemaking and insight is not easy. Many data mining and machine learning algorithms are considered black-boxes, meaning that they are difficult to understand, which hinders their effective use and makes their results difficult to trust. How to make the models transparent and easy to use remains a great challenge within the VA community.
- Trustworthiness and provenance: Not only the models used in the VA process need to be transparent, but also the data used and refined. For example, uncertainty may arise and spread in different steps of an analytics process and to keep track of the uncertainty is important for producing reliable and trustworthy results. Many techniques for uncertainty visualization exist, but due to the complexity of different VA applications, there are still no widely accepted techniques. Moreover, to enable the analyst to keep track of the VA process and the intermediate results is important, especially in collaborative scenarios, to promote understanding and transparency. However, more research is needed how to retrieve and visualize important partial results.
- Tool design and evaluation: VA has been applied in a wide range of domains and many application-specific tools have been developed. Due to the heterogeneity of the applications and users, it is not easy to accommodate for all needs and preferences. For example, experts may require flexibility, whereas novices guidance regarding appropriate analysis tools and visualizations for the task at hand. VA practitioners have used various approaches such as case studies, expert reviews, formal/informal user studies etc. to evaluate the usability and effectiveness of the systems, and more guidance is needed how to perform good evaluations.

8.5 Design and Evaluation

This section presents basic guidelines and processes for the design and evaluation of InfoVis and VA techniques, methods and systems.

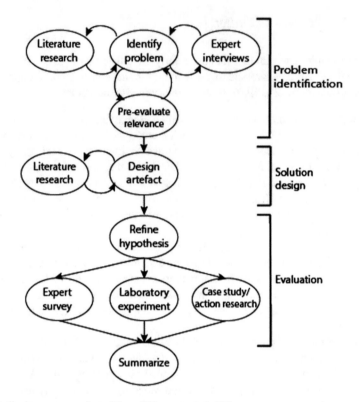

Fig. 8.6 Design process adapted from Offermann et al. [47]

8.5.1 Design

The design space of possible ways to design visualization systems and techniques is vast [44]. There are design processes in the areas of system design and user-centered design (that involve capturing the user's needs and preferences) that can be used for InfoVis and VA design, for instance, Vredenburg et al. [67] and Offermann et al. [47]. As shown in Fig. 8.6, the process established the following phases: identify the problem, pre-evaluate relevance, design artifact, refine hypothesis, expert survey/laboratory experiment/case study/action research and summarise. A practical example of the application of this design process can be found in [51].

A design that does well by one measure might rate poorly on another, hence, the characterization of trade-offs in the visualization design space is an open problem in visualization research [44]. A suggested process for the design and validation of visualizations is the one presented by Munzner in [43], see Figs. 8.7 and 8.8. This model splits visualization design into levels; the four levels are: characterize the tasks and data in the vocabulary of the problem domain, abstract into operations and data types, design visual encoding and interaction techniques, and create algorithms to

Fig. 8.7 Four levels of the nested model and four kinds of threats to validity adapted from Munzner [43]

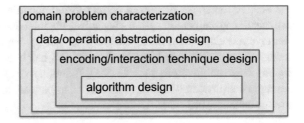

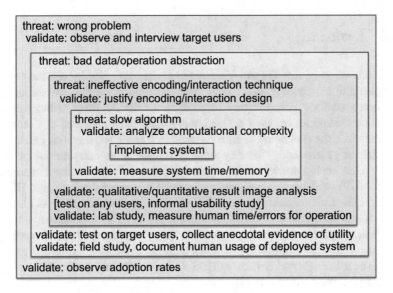

Fig. 8.8 Nested model for visualization design and validation, adapted from Munzner [43]

execute these techniques efficiently. Munzner [43] shows four kinds of threats to validity:

- wrong problem (they do not do that),
- wrong abstraction (you are showing them the wrong thing),
- wrong encoding/interaction technique (the way you show it does not work),
- wrong algorithm (your code is too slow).

The top level characterizes the problems and data of a particular domain, the next level maps those into abstract operations and data types, the third level is to design the visual encoding and interaction to support those operations, and the innermost fourth level is to create an algorithm to carry out that design automatically and efficiently. The three inner levels are all instances of design problems, although it is a different problem at each level. These levels are nested, so the output from an upstream level above is input to the downstream level below, hence an upstream error inevitably cascades to all downstream levels. That is, if a poor choice was made in the abstraction stage, then even perfect visual encoding and algorithm design will not create a visualization system that solves the intended problem. The model shows iterative refinement and the levels don't need to be done in strict order. The InfoVis design process illustrated in Figs. 8.7 and 8.8 has been revisited and improved in several later publications [41, 42, 55]. Examples of design papers from our own research are presented in [50, 51]; other examples of InfoVis and VA design papers are for instance, [40, 45].

8.5.2   Evaluation

The proliferation of InfoVis and VA techniques has also highlighted the need for principles and methodologies for their empirical evaluation [17]. Research in information visualization has largely focused on the development of innovative visualization techniques but their evaluation has often been relegated to a secondary role. However, the evaluation of InfoVis and VA techniques, methods and systems is crucial for guaranteeing that they actually support users carrying out analytical tasks. Evaluation is a crucial part of the design process. When talking about evaluation of InfoVis and VA is necessary that we refer to other disciplines, e.g. HCI and *usability*. The international standard, ISO 9241-11, defines *usability* as the extent to which a product can be used by *specified users* to achieve *specified goals* with *effectiveness, efficiency,* and *satisfaction* in a *specified context of use*. According to its purpose, there are three general types of usability evaluation: *exploratory* (how is it or will be used?), *formative* (how can it be made better?) and *summative* (how good is it?) [1] (see notes by Andrews [2]).

   Exploratory evaluation examines current usage and the potential design space for new designs. It is normally carried out before the interface design; learning *which* software is used, *how often* and *what for*. Normally, *usage data* is collected— statistical summaries and observations of usage.

   → Use observation, interviews, surveys and automated logging.

   Formative evaluation informs the design process and helps improving and interface during the design process. It learns why something went wrong, not just that it went wrong. *Process data* is collected during this kind of evaluation—qualitative observations of what happened and why.

   → Heuristic evaluations and thinking aloud tests should be run at regular intervals.

   Summative evaluation assess the overall quality of an interface. It is done once an interface is more or less *finished*. Either compare alternative designs, or tests definite performance requirements. It collects *bottom-line data*—quantitative *measurements*

---

[1] Soup analogy:
"When the cook tastes other cook's soups, that's exploratory.
When the cook tastes his own soup while making it, that's formative.
When the guests (or food critics) taste the soup, that's summative."

of performance: how long did users take, were they successful, how many errors did they make.

→ Running formal experiments that measure, for example: time to complete specific tasks, number of tasks completed within a given time, accuracy of task completion, number of errors, number of commands/features used.

There are three challenging aspects when evaluating InfoVis and VA techniques and systems that deserve special consideration: how to measure *insight*, evaluation *tasks* and evaluation *metrics*.

Insight. Providing insight is considered to be the main purpose of InfoVis and VA [46, 61]. However, there is no comprehensive evaluation methodology to determine the real value of visualization in terms of its goals of facilitating insight, discovering knowledge, or supporting analytical reasoning. Part of the problem of designing and developing adequate evaluation methodologies resides in the fact that it is difficult to define insight or knowledge discovery, and even if definitions exist (see an exhaustive discussion in [15], there is not one that is commonly accepted by the research community [71]. How do the newly designed visual environments assist users in gaining insight, acquiring knowledge, or reasoning analytically?

The approaches taken to study insight and analytical reasoning stem from cognitive acience theories, sense-making, activity theory, problem solving or human computer interaction. For instance, in [60], the authors demonstrate that *cognitive fit theory*, along with the proximity compatibility principle, can be used as a basis to evaluate the effectiveness of information visualizations to support a decision-making task. The cognitive fit theory proposes that when the problem representation fits the problem-solving task, a preferable mental representation of the problem will be created, resulting in improving the accuracy and speed of the problem-solving process [60].

Evaluation tasks. The lack of task models and taxonomies of tasks makes the assessment of visual environments for researchers and developers that are not experienced in evaluation more difficult. Which tasks should be used when evaluating InfoVis and VA environments?

There are several classifications of tasks for evaluation purposes in the literature, a recent review that covers such classifications and a discussion of their terminology, nature and validity is presented in [31], while an example classification with particular tasks to use during the evaluation can be found in [12, 54]. For instance, a visual task can be 'to categorize', 'to distinguish', 'to compare', 'to identify', 'to locate', 'to rank', 'to cluster', 'to associate', etc.

Nevertheless, Saraiya et al. [52] argue that most visualization tools are evaluated in short-term and controlled studies, using preselected data sets and benchmark tasks. Therefore, in [52], the authors present a long-term study of the use of certain VA tools (such as Spotfire, PathwayAssist and GenMapp) during the analysis of a bioinformatics data set. To keep the experiment as close to a real-world data analysis as possible, no predefined tasks were used, and the bioinformaticians were requested to keep a diary with the insights gained from the data analyzed. One of the conclusions drawn is that *longitudinal studies* can provide insight into the VA process, practices, and actual data analysis tasks that can guide evaluators and visualization designers in constructing tools that better match this analytic process.

Evaluation metrics. The usage and definition of quality metrics for visualization techniques are, as yet, immature fields and there is no common understanding of metrics [10]. Bertini and Santucci [10] propose a classification for visual metrics based on three main categories: size metrics (e.g., number of data items, density, and screen occupation percentage), visual effectiveness metrics (measures of occlusion, collisions, and outliers), and features preservation metrics (intended for measuring how correctly an image represents some data characteristic, e.g., number of identifiable points compared to number of actual data items).

Papers dealing with InfoVis and VA evaluation that can be used as examples for how to carry out this kind of work are [25, 48, 49].

References

1. Abadi, M., Agarwal, A., Barham, P., Brevdo, E., Chen, Z., Citro, C., Corrado, G. S., et al. (2015). TensorFlow: Large-scale machine learning on heterogeneous systems. https://www.tensorflow.org/. Software available from tensorflow.org.
2. Andrews, K. (2010). Usability evaluation and information visualisation. IV'10 tutorial. In *14th International Conference Information Visualisation, London, UK.*
3. Bae, J., Helldin, T., & Riveiro, M. (2017). Understanding indirect causal relationships in node-link graphs. In *Computer graphics forum.* https://doi.org/10.1111/cgf.13198.
4. Bae, J., & Watson, B. (2011). Developing and evaluating quilts for the depiction of large layered graphs. *IEEE Transactions on Visualization & Computer Graphics, 17,* 2268–2275. https://doi.org/10.1109/TVCG.2011.187.
5. Bastian, M., Heymann, S., & Jacomy, M. (2009). Gephi: An open source software for exploring and manipulating networks. http://www.aaai.org/ocs/index.php/ICWSM/09/paper/view/154.
6. Bederson, B. B., & Shneiderman, B. (2003). In: *The craft of information visualization, interactive technologies.* San Francisco: Morgan kaufmann. https://doi.org/10.1016/B978-1-55860-915-0.50056-1. https://www.sciencedirect.com/science/article/pii/B9781558609150500561.
7. Belmonte, N.G. (2011). *JavaScript InfoVis Toolkit.* Retrieved October 7, 2017, from http://philogb.github.io/jit/index.html.
8. Bertin, J. (1983). *Semiology of graphics.* University of Wisconsin Press.
9. Bertin, J. (2011). *Semiology of graphics: Diagrams, networks, maps.* Esri Press.
10. Bertini, E., & Santucci, G. (2006). Visual quality metrics. In *Proceedings of the 2006 AVI Workshop on Beyond Time and Errors: Novel Evaluation Methods for Information Visualization* (pp. 1–5). ACM.
11. Bostock, M., Ogievetsky, V., & Heer, J. (2011). D3: Data-driven documents. *IEEE Transactions on Visualization and Computer Graphics (Proc. InfoVis).* http://vis.stanford.edu/papers/d3.
12. Brehmer, M., & Munzner, T. (2013). A multi-level typology of abstract visualization tasks. *IEEE Transactions on Visualization and Computer Graphics, 19*(12), 2376–2385.
13. Card, S., Mackinlay, J. D., & Shneiderman, B. (1999). *Readings in information visualization: Using vision to think.* London: Academic Press.
14. Castellanos-Garzón, J. A., García, C. A., Novais, P., & Díaz, F. (2013). A visual analytics framework for cluster analysis of dna microarray data. *Expert Systems with Applications, 40*(2), 758–774.
15. Chang, R., Ziemkiewicz, C., Green, T. M., & Ribarsky, W. (2009). Defining insight for visual analytics. *IEEE Computer Graphics and Applications, 29*(2), 14–17.

16. Chen, C. (2005). Top 10 unsolved information visualization problems. *IEEE Computer Graphics and Applications, 25*(4), 12–16. https://doi.org/10.1109/MCG.2005.91. http://dx.doi.org/10.1109/MCG.2005.91.
17. Chen, C., & Czerwinski, M. (2000). Empirical evaluation of information visualizations: An introduction. *International Journal of Human-Computer Studies, 53*(5), 631–635.
18. Endert, A., Ribarsky, W., Turkay, C., Wong, B., Nabney, I., Daz Blanco, I., et al. (2017). The state of the art in integrating machine learning into visual analytics. https://doi.org/10.1111/cgf.13092.
19. Few, S. (2009). *Now you see it: Simple visualization techniques for quantitative analysis.* Oakland: Analytics Press.
20. Goodall, J. R., & Tesone, D. R. (2009). Visual analytics for network flow analysis. In *Conference For Homeland Security, 2009. CATCH 2009. Cybersecurity Applications & Technology* (pp. 199–204). IEEE.
21. Purchase, H. C., Andrienko, T, N., Jankun-Kelly, J., & Ward, M. (2008). *Theoretical foundations of information visualization* (Vol. 4950). Lecture notes in computer science. Berlin: Springer.
22. Heer, J., Bostock, M., & Ogievetsky, V. (2010). A tour through the visualization zoo. *Communications of the ACM, 53*(6), 59–67. https://doi.org/10.1145/1743546.1743567. http://doi.acm.org/10.1145/1743546.1743567.
23. Heer, J., Card, S. K., & Landay, J. (2005) Prefuse: A toolkit for interactive information visualization. In *ACM human factors in computing systems (CHI)* (pp. 421–430). http://vis.stanford.edu/papers/prefuse.
24. Inc., P.T. (2015). Collaborative data science. https://plot.ly.
25. Kay, M., & Heer, J. (2016). Beyond weber's law: A second look at ranking visualizations of correlation. *IEEE Transactions on Visualization and Computer Graphics, 22*(1), 469–478.
26. Keim, D., Kohlhammer, J., Ellis, G., & Mansmann, F. (2010). Mastering the information age: Solving problems with visual analytics. In *Eurographics* (Vol. 2, p. 5).
27. Keim, D., Mansmann, F., Oelke, D., & Ziegler, H. (2008) Visual analytics: Combining automated discovery with interactive visualizations. In *Discovery science* (pp. 2–14). Springer.
28. Keim, D. A. (2002). Information visualization and visual data mining. *IEEE Transactions on Visualization and Computer Graphics, 8*(1), 1–8.
29. Keim, D. A., Mansmann, F., Schneidewind, J., & Ziegler, H. (2006). Challenges in visual data analysis. In *Proceedings of the 10th International Conference on Information Visualization* (pp. 9–16). IEEE.
30. Keim, D. A., Munzner, T., Rossi, F., & Verleysen, M. (2015). Bridging information visualization with machine learning (Dagstuhl Seminar 15101). *Dagstuhl Reports, 5*(3), 1–27. https://doi.org/10.4230/DagRep.5.3.1. http://drops.dagstuhl.de/opus/volltexte/2015/5266.
31. Kerracher, N., & Kennedy, J. (2017). Constructing and evaluating visualisation task classifications: Process and considerations. In *Computer graphics forum* (Vol. 36, pp. 47–59). Wiley Online Library.
32. Kerren, A., & Schreiber, F. (2012). Toward the role of interaction in visual analytics. In *Proceedings of the 2012 Winter Simulation Conference (WSC)* (pp. 1–13). IEEE.
33. Kiln.it (2014). *In flight.* Retrieved October 7, 2017, from https://www.theguardian.com/world/ng-interactive/2014/aviation-100-years.
34. Kosara, R. (2016). An empire built on sand: Reexamining what we think we know about visualization. In *Proceedings of the Sixth Workshop on Beyond Time and Errors on Novel Evaluation Methods for Visualization, BELIV 2016* (pp. 162–168). New York, USA: ACM. https://doi.org/10.1145/2993901.2993909. http://doi.acm.org/10.1145/2993901.2993909.
35. Liu, S., Cui, W., Wu, Y., & Liu, M. (2014). A survey on information visualization: Recent advances and challenges. *The Visual Computer, 30*(12), 1373–1393. https://doi.org/10.1007/s00371-013-0892-3. http://dx.doi.org/10.1007/s00371-013-0892-3.
36. Mansmann, F. (2008). *Visual analysis of network traffic: Interactive monitoring, detection, and interpretation of security threats.* Ph.D. thesis.
37. Mark, G., & Kobsa, A. (2005). The effects of collaboration and system transparency on cive usage: An empirical study and model. *Presence: Teleoperators and Virtual Environments, 14*(1), 60–80. https://doi.org/10.1162/1054746053890279.

38. Mauri, M., Elli, T., Caviglia, G., Uboldi, G., & Azzi, M. (2017). Rawgraphs: A visualisation platform to create open outputs. In *Proceedings of the 12th Biannual Conference on Italian SIGCHI Chapter, CHItaly 2017* (pp. 28:1–28:5). ACM. https://doi.org/10.1145/3125571. 3125585. http://doi.acm.org/10.1145/3125571.3125585.
39. McGuirl, J. M., & Sarter, N. B. (2006). Supporting trust calibration and the effective use of decision aids by presenting dynamic system confidence information. *Human Factors, 48*(4), 656–665. https://doi.org/10.1518/001872006779166334. PMID: 17240714.
40. Meyer, M., Munzner, T., & Pfister, H. (2009). Mizbee: A multiscale synteny browser. *IEEE Transactions on Visualization and Computer Graphics, 15*(6), 897–904.
41. Meyer, M., Sedlmair, M., & Munzner, T. (2012). The four-level nested model revisited: Blocks and guidelines. In *Proceedings of the 2012 BELIV Workshop: Beyond Time and Errors-Novel Evaluation Methods for Visualization* (p. 11). ACM.
42. Meyer, M., Sedlmair, M., Quinan, P. S., & Munzner, T. (2015). The nested blocks and guidelines model. *Information Visualization, 14*(3), 234–249.
43. Munzner, T. (2009). A nested model for visualization design and validation. *IEEE Transactions on Visualization and Computer Graphics, 15*(6).
44. Munzner, T. (2014). *Visualization analysis and design*. CRC Press.
45. Nielsen, C. B., Jackman, S. D., Birol, I., & Jones, S. J. (2009). Abyss-explorer: Visualizing genome sequence assemblies. *IEEE Transactions on Visualization and Computer Graphics, 15*(6), 881–888.
46. North, C. (2006). Toward measuring visualization insight. In *Position paper for the IEEE VAST Metrics for the Evaluation of Visual Analytics Workshop*.
47. Offermann, P., Levina, O., Schönherr, M., & Bub, U. (2009). Outline of a design science research process. In *Proceedings of the 4th International Conference on Design Science Research in Information Systems and Technology* (pp. 1–11). New York, USA: ACM. http://doi.acm.org/10.1145/1555619.1555629.
48. Riveiro, M. (2014). Evaluation of normal model visualization for anomaly detection in maritime traffic. *ACM Transactions on Interactive Intelligent Systems (TiiS), 4*(1), 5.
49. Riveiro, M., Helldin, T., Falkman, G., & Lebram, M. (2014). Effects of visualizing uncertainty on decision-making in a target identification scenario. *Computers & Graphics, 41*, 84–98.
50. Riveiro, M., Helldin, T., Lebram, M., & Falkman, G. (2013). Towards future threat evaluation systems: User study, proposal and precepts for design. In *2013 16th International Conference on Information Fusion (FUSION)* (pp. 1863–1870). IEEE.
51. Riveiro, M., Lebram, M., & Warston, H. (2014). On visualizing threat evaluation configuration processes: A design proposal. In *2014 17th International Conference on Information Fusion (FUSION)* (pp. 1–8). IEEE.
52. Saraiya, P., North, C., Lam, V., & Duca, K. A. (2006). An insight-based longitudinal study of visual analytics. *IEEE Transactions on Visualization and Computer Graphics, 12*(6), 1511–1522.
53. Satyanarayan, A., Moritz, D., Wongsuphasawat, K., & Heer, J. (2017). Vega-lite: A grammar of interactive graphics. *IEEE Transactions on Visualization and Computer Graphics (Proc. InfoVis)*. http://idl.cs.washington.edu/papers/vega-lite.
54. Schulz, H. J., Nocke, T., Heitzler, M., & Schumann, H. (2013). A design space of visualization tasks. *IEEE Transactions on Visualization and Computer Graphics, 19*(12), 2366–2375.
55. Sedlmair, M., Meyer, M., & Munzner, T. (2012). Design study methodology: Reflections from the trenches and the stacks. *IEEE Transactions on Visualization and Computer graphics, 18*(12), 2431–2440.
56. Shneiderman, B. (1992). Tree visualization with tree-maps: 2-d space-filling approach. *ACM Transactions on Graphics, 11*(1), 92–99. http://doi.acm.org/10.1145/102377.115768.
57. Shneiderman, B. (1996). The eyes have it: A task by data type taxonomy for information visualizations. In *Proceedings of the IEEE Symposium on Visual Languages* (pp. 336–343). IEEE.
58. Slingsby, A., Dykes, J., & Wood, J. (2011). Exploring uncertainty in geodemographics with interactive graphics. *IEEE Transactions on Visualization and Computer Graphics, 17*(12), 2545–2554.

59. Sun, G. D., Wu, Y. C., Liang, R. H., & Liu, S. X. (2013). A survey of visual analytics techniques and applications: State-of-the-art research and future challenges. *Journal of Computer Science and Technology, 28*(5), 852–867.

60. Teets, J. M., Tegarden, D. P., & Russell, R. S. (2010). Using cognitive fit theory to evaluate the effectiveness of information visualizations: An example using quality assurance data. *IEEE Transactions on Visualization and Computer Graphics, 16*(5), 841–853.

61. Thomas, J., & Cook, K. (2005). *Illuminating the path: The research and development agenda for visual analytics*. IEEE Computer Society Press.

62. Thomas, J., & Cook, K. (2006). A visual analytics agenda. *Computer Graphics and Applications, 26*(1), 10–13.

63. Thomas, J., & Kielman, J. (2009). Challenges for visual analytics. *Information Visualization, 8*(4), 309–314.

64. Treisman, A., & Gormican, S. (1988). Feature analysis in early vision: Evidence from search asymmetries. *Psychological Review, 95*(1), 15–48.

65. Tufte, E. (2001). *The visual display of quantitative information*. Cheshire, Conn: Graphics Press.

66. van der Maaten, L., & Hinton, G. (2008). Visualizing high-dimensional data using t-SNE. *Journal of Machine Learning Research, 9*, 2579–2605.

67. Vredenburg, K., Mao, J. Y., Smith, P. W., & Carey, T. (2002). A survey of user-centered design practice. In *Proceedings of the SIGCHI Conference on Human Factors in Computing Systems, CHI 2002* (pp. 471–478). New York, USA: ACM.

68. Ward, M., Grinstein, G., & Keim, D. (2010). *Interactive data visualization: Foundations, techniques, and applications. 360 degree business*. CRC Press.

69. Ware, C. (2008). *Visual thinking for design* (1st ed.). Oxford: Elsevier LTD.

70. Ware, C. (2012). *Information visualization: Perception for design* (3rd ed.). Oxford: Elsevier LTD.

71. Yi, J. S., Kang, Y. a., Stasko, J. T., & Jacko, J. A. (2008). Understanding and characterizing insights: How do people gain insights using information visualization? In *Proceedings of the 2008 Workshop on Beyond Time and Errors: Novel Evaluation Methods for Information Visualization* (p. 4). ACM.

Chapter 9
Complex Data Analysis

Juhee Bae, Alexander Karlsson, Jonas Mellin, Niclas Ståhl and Vicenç Torra

Abstract Data science applications often need to deal with data that does not fit into the standard entity-attribute-value model. In this chapter we discuss three of these other types of data. We discuss texts, images and graphs. The importance of social media is one of the reason for the interest on graphs as they are a way to represent social networks and, in general, any type of interaction between people. In this chapter we present examples of tools that can be used to extract information and, thus, analyze these three types of data. In particular, we discuss topic modeling using a hierarchical statistical model as a way to extract relevant topics from texts, image analysis using convolutional neural networks, and measures and visual methods to summarize information from graphs.

9.1 Introduction

Traditional databases consist of tables where data is represented in terms of the entity-attribute-value model and has a rigid structure. That is, data is represented in terms of a set of records where each record describes an entity by means of values for each attribute. Nevertheless, there is information that can be represented in an easier way in other formats. NoSQL databases, for *not only SQL*, permit efficient storage and retrieval of other types of data. This is the case of text documents and

J. Bae (✉) · A. Karlsson · J. Mellin · N. Ståhl · V. Torra
School of Informatics, University of Skövde, Skövde, Sweden
e-mail: juhee.bae@his.se

A. Karlsson
e-mail: alexander.karlsson@his.se

J. Mellin
e-mail: jonas.mellin@his.se

N. Ståhl
e-mail: niclas.stahl@his.se

V. Torra
e-mail: vtorra@his.se; vtorra@ieee.org

© Springer International Publishing AG, part of Springer Nature 2019   157
A. Said and V. Torra (eds.), *Data Science in Practice*, Studies in Big Data 46,
https://doi.org/10.1007/978-3-319-97556-6_9

images, key-value objects (collection in which different objects may have different sets of fields) and graphs.

Big data introduces additional challenges to storage, processing and analysis. Big data is usually defined in terms of the well known 3Vs: Volume (data in large quantities), Variety (from text to images, geolocations and all kinds of logs), Velocity (data comes fast and needs to be processed also fast, and sometimes data life time is short). Sometimes additional sets of Vs (5Vs or even 7Vs) are used to define big data. 5Vs include Veracity (trustworthiness, quality and accuracy of the data) and Value (as the goal is to turn data into value). 7Vs include Variability (data constantly changing) and Visualization (to stress the need of tools to help understanding the data). N.B., we are not addressing the legal and organizational sources of complexity that are prevalent in the public sector [8, 15] where two S's and 1 V are considered: Silos, Security, and Variety.

Taking all this into account and in order to be operational, we can distinguish three main subclasses within big data: (i) large volumes, (ii) streaming data, and (iii) dynamic data. Large volumes refers to data that is of large volume but with a low variability. Streaming data is when data arrives continuously and we need to process it in real time. Dynamic data is when data continuously changes. Methods and algorithms for data storage and processing are defined taking into account these types of data.

When we take a snap-shot of a part of a social network for its analysis, we have big data in the sense of (i). That is, we have data of large volume but without variability. On the contrary, when we are interested in posts in a social network and how their content changes with respect to time, we are stressing variability.

Methods for streaming data, (ii) above, are often based on sliding windows (see e.g., [5]). That is, data is buffered and processed in memory. It is clear that the size of the window influences the quality of the processing. Larger windows usually improve the quality of the data processing at the cost of computational time and memory. Adaptive windowing [5] has also been proposed to improve results.

Dynamic data, (iii) above, means that the result of any analysis change over time. We can repeat the analysis or revise previous results and conclusions.

In this chapter we focus on the analysis of data of type (i). We focus on three types of data and discuss some analyses for these types. In particular, we present examples of tools to analyze text documents, images and graphs.

It is important to stress that in the analysis of complex data, the selection of the problem and the methods to be used are of crucial importance. They are in fact more relevant than in the analysis of non-complex data. This is so because there are more standardized methods for the analysis and for sketching the latter type of data. Another important issue for any data scientist is whether complex data can be transformed into a non-complex data set to apply standard methods. For example, extracting features. Transformation of text documents into bags of words and then apply a machine learning method to these bags of words is an example of this type of transformation.

The structure of this chapter is as follows. We will begin in Sect. 9.2 showing how to build a hierarchical statistical model to find the topics of a text. In Sect. 9.3 we consider image analysis using convolutional neural networks. Then, in Sect. 9.4 we discuss graphs and discuss some of the measures used to extract information from them, and also the use of quilts for their visualization. The chapter finishes with a discussion and summary.

9.2 Text Analysis and Topic Modeling

Topic models [6, 7] is a modeling technique used for text analysis where the aim is to capture the contents of the text in the form of patterns in word distributions of the documents, denoted as *topics*. The method constitutes an unsupervised learning approach to text analysis. When a topic model is a good "fit" to a given *corpus*, i.e., set of documents, and if one possesses domain knowledge regarding the text, it is often possible to label the topics by exploring the words found in the topics. As an example, assuming that a topic model has been trained on a corpus about programming, one might find a topic with the top five most probable words: *divide, dynamic, conquer, algorithm, complexity*, which one could then label as "design paradigms of algorithms". Note that there is no mentioning of any words like "design" or "paradigm" in the above words, nevertheless, by using domain knowledge, the label of the topic can still be inferred.

More formally, a topic model is a *hierarchical statistical model* where the topics, i.e., word distributions $T = \{p(w|t_1), \ldots, p(w|t_n)\}$, reside on the top level within the model, i.e., the corpus level D, where each document $D \in D$ consists of words w, according to some pre-defined vocabulary of words V. This means that any document $D \in D$ is reduced to the words found in V and that each such word is assumed to originate, or be *generated*, from a topic found in T. Hence, one can think of a document as having a proportion of each topic. An illustration of such line of thinking can be seen in Fig. 9.1.

In order to define a vocabulary V, it is necessary to perform some pre-processing on the documents. We reduce each word into its "base form" (so different versions of the same word are not counted as different words) denoted as *stemming*, and remove so called stop words (words like "the" etc. that do not provide much information), numbers and short words [1, 13]. After this one constructs a so called *document-term* matrix that contains the frequencies of each word type. This matrix can be furthered used to filter out important words according to a weighting schema, e.g., *term-frequency inverse-document frequency* [18].

In order to learn a topic model, one needs to resort to *approximate inference* algorithms, e.g., *Gibbs sampling* [12]. An illustration of the learning phase of the topic model is seen in Fig. 9.2.

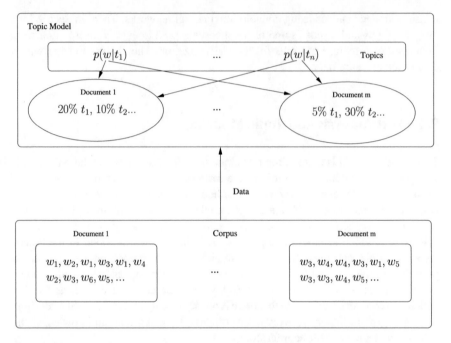

Fig. 9.1 The figure shows that each document is transformed to a set of words (bottom part) according to a vocabulary $\mathcal{V}$ and then this information is used in order to learn a topic model, shown in the upper half section of the figure. We can see that the words in each document are modeled as if they were originating from a proportion of different topics on the corpus level

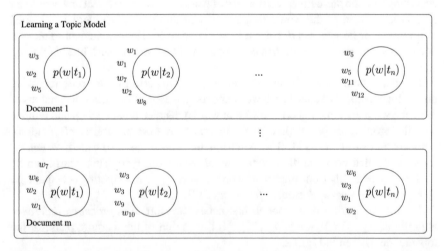

Fig. 9.2 The figure illustrates that words have been assigned to topics and by iterating through each word w and sample from the conditional distribution given the word-to-topic assignment for all other words, besides w, one can eventually obtain a sample that is representative for the topic model

9.3 Image Analysis with Convolutional Neural Networks

Computer vision or *machine vision* is one of the oldest and most studied fields within AI [21]. The aim of *computer vision* is to mimic the human visual system and let computers extract high-level understanding from images. Even if computer vision systems have been developed a lot in the last decades many challenges are still to be solved. A major problem is to understand the contextual structure of the image, that is, to understand the relationships and interactions among the objects in the image.

In this section, we will look at more specific image analysis tasks in which computers have excelled, such as *object detection* in images [19] and *image classification* [16]. One of the reasons for why image analysis has advanced so much in the last decades is due to the introduction of convolutional neural networks. Yet much of the advancements are also driven by practical improvements such as the vast increase in available data and the increase of computational performance. In this section, we describe how to perform an image analysis using convolutional neural networks. See also the chapter on machine learning in this book [10].

Even if it is not strictly required in all uses for convolutional neural networks, it is often a appropriate to scale or crop images so that all images in the dataset have the same shape and number of pixels.

The idea behind convolutional neural networks is that each small area in an image may be analyzed separately and in the same way. This analysis, called the convolutional step, finds hidden features describing the content of the given small area. The found hidden features may then be combined into matrices, called feature maps, that can be analyzed in the same way as the image was analyzed. When the hidden features are combined it is common to use some kind of subsampling, such as average or max pooling, representing an area with either its average or maximum value. This is called the pooling step. Convolutional and pooling steps are the main building blocks of a *convolutional neural network* and will be described with greater detail in this section. Finally, it will be described how these are combined into a *convolutional neural network*.

9.3.1 Convolutional Step

In the convolutional step the image or map is split up in several small squares, often arranged in a grid and overlapping each other (see Fig. 9.3). A single layered neural network is applied to these small areas and the output is defined as:

$$f\left(\theta_0 + \sum_{i=0}^{m}\sum_{j=0}^{n}\theta_{i,j}x_i\right) \qquad (9.1)$$

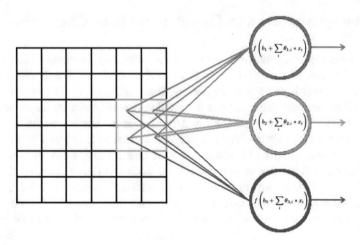

Fig. 9.3 A square of size 2×2 is being analyzed by a network with three neurons. The square that are analyzed are highlighted. All possible square of size 2×2 in the image will then be analyzed separately, but by the same network, having the same weights

where $\theta_{i,j}$ are the weight parameters that should be learned by the network, θ_0 is the bias that also should be learned and f is the activation function. The rectified linear function defined in Eq. (9.2) is often used as activation function [9].

$$f_{relu}(x) = \begin{cases} x & \text{if } x > 0 \\ 0 & \text{otherwise} \end{cases} \tag{9.2}$$

In Eq. (9.1) m represent the number of columns in the small square and n the number of rows. An illustration of this is shown in Fig. 9.3.

To further illustrate this process the results of applying convolutional neural networks with different weights to all possible squares of size 6×6 pixels of the image shown in Fig. 9.4a is shown in Fig. 9.4b and c. The first network that we apply to this image is designed to recognize a horizontal transformation from dark to light. The weights of this network is defined as:

$$\theta^{(1)} = \begin{pmatrix} 1 & 1 & 1 & -1 & -1 & -1 \\ 1 & 1 & 1 & -1 & -1 & -1 \\ 1 & 1 & 1 & -1 & -1 & -1 \\ 1 & 1 & 1 & -1 & -1 & -1 \\ 1 & 1 & 1 & -1 & -1 & -1 \\ 1 & 1 & 1 & -1 & -1 & -1 \end{pmatrix} \tag{9.3}$$

The second network is designed to detect the opposite transformation and is thus defined as $\theta^{(2)} = -\theta^{(1)}$. The result of applying the first network is shown in Fig. 9.4b and the result from the second network is shown in Fig. 9.4c. In both examples the

(a) The original image that are being analyzed.

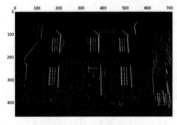

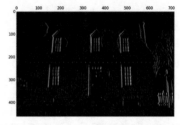

(b) The resulting feature map of a convolutional neural network with the weights $\theta^{(1)}$, applied to every 6×6 square of the image in Figure 9.4a.

(c) The resulting feature map of a convolutional neural network with the weights $\theta^{(2)}$, applied to every 6×6 square of the image in Figure 9.4a.

Fig. 9.4 The output of a convolutional neural network with two different weight settings applied to the image in Fig. 9.4a. The weights $\theta^{(1)}$ are defined in Eq. (9.3) and $\theta^{(2)}$ is defined as $-\theta^{(1)}$

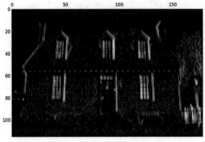

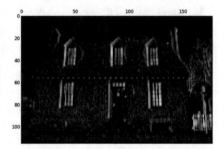

(a) Max pooling applied to Figure 9.4a.

(b) Max pooling applied to Figure 9.4b.

Fig. 9.5 Max pooling applied to the two features maps shown in Fig. 9.4

bias (θ_0) is equal to zero. This process could be equated with applying the filter to each small square.

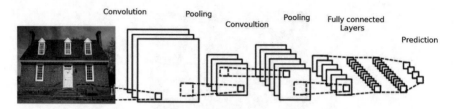

Fig. 9.6 A combination of convolutional and pooling layers forming a convolutional neural network. Note that the last layers are fully connected feed forward layers

9.3.2 Pooling Step

In the pooling step the dimension of the feature maps is reduced, while the important information is preserved. Several ways of pooling can be used, such as average pooling or max pooling, where max pooling is the most commonly used. When applying max pooling only the maximum value of a given neighborhood is kept. The result of applying maximum pooling on 4×4 squares of the images in Fig. 9.4b and c is shown in Fig. 9.5.

9.3.3 Putting It All Together

A full convolutional neural network is built up by combining convolutional steps and pooling steps. It is also common to have some fully connected networks layers at the end of the convolutional neural network. An example of the layout of a convolutional neural network is shown in Fig. 9.6.

In the examples above all values are hand crafted to give an intuitive understanding of how a convolutional neural network works. This is not the case in reality where the weights are learned by minimizing a given cost function over a data set with input values and target values or classes. However, it has been shown that a convolutional neural network often learn to detect the same things as common hand crafted filters, such as edge detection [25].

9.4 Graph Analysis

Graphs can be used to model a large number of systems. They permit to represent the interaction between objects. Social networks are a typical example of systems that can be represented in terms of graphs. In this case, the nodes represent the people and the edges represent the relationships between the people. Communication networks (roads or internet) are another example. Therefore, we need to explore graphs to understand the relationship between the objects.

When a graph consists only of a few nodes, we can represent it using a node-link diagram and then understand the relationships between the nodes by visual inspection. In general, however, this is not possible. In the era of big data, and when social networks have millions of users, this type of visualization is naturally out of the question. Another way to analyze a graph is visualizing the adjacency matrix. Similar problems arise with large graphs. In addition, current graphs include many attributes and features (both related to the nodes and to the edges).

Because of that a few approaches have been considered for graph analysis. On the one hand a few measures have been defined to give summaries of the information in a graph. These are, for example, degree, centrality, and clustering coefficients. On the other hand visual methods have been developed to see graphically the structure and the properties (e.g., attributes for nodes and edges) of the graph.

Popular multi-dimensional visualizations include parallel coordinates [14] and scatterplot matrices [11]. Researchers are now integrating these visualizations together, or with network diagrams so that people can interpret the relations in various perspectives.

9.4.1 Measures

Let $G = (V, E)$ denote a graph in terms of the set of nodes V and the set of edges E. As edges connect two nodes, it is clear that $E \subseteq V \times V$. A path is a set of edges that connect two nodes. We will use $d(u, v)$ to denote the distance of the shortest path between nodes u and v. Note that the shortest path is also called geodesic.

Several measures have been proposed in the literature for graphs. Some of them are defined for the graph itself (e.g., the graph diameter, the girth), others are defined for nodes (e.g., the degree, the eccentricity, and most centrality measures), or for the edges (e.g., the edge betweenness centrality).

The following measures are defined for nodes:

- Eccentricity, denoted by $\epsilon(v)$, is the greatest distance between v and any other node in the graph. That is, $\epsilon(v) = \max_{u \in V} d(u, v)$.
- Degree, denoted by $\delta(v)$, is the number of edges that connect v with any other node.
- Centrality measures are to find those nodes in the graph that are influential or important in the graph. The meaning of influential and importance, of course, is application dependent, and that is why, several of such measures have been defined. Degree centrality, betweenness, and closeness are examples of such measures:

  - Degree centrality of a node corresponds to the degree of the node.
  - Closeness of a node is the reciprocal of the average shortest path from the given node. That is,

$$c(u) = \left(\frac{\sum_{v \in V} d(u, v)}{|V| - 1} \right)^{-1}.$$

In this way, when all nodes are at a distance 1, the average shortest path is 1 and the $c(u)$ is also 1. Otherwise, when there are very long shortest paths, the average will be large and the closeness can be near to zero.

- Betweenness of a node counts the number of shortest paths that go through the node. Let $\sigma_{nm}(v)$ be the number of shortest paths from node n to node m through v, and σ_{nm} the number of shortest paths from n to m. Then, betweenness is defined by

$$B(u) = \sum_{n \in V, m \in V, n \neq u \neq m} \frac{\sigma_{nm}(v)}{\sigma_{nm}}.$$

The concept of betweenness centrality is defined for edges in a similar way to the one for edges.

- Edge betweenness centrality for an edge is defined as the number of shortest paths that go through the edge.

Finally, we review some other measures for graphs.

- The diameter of a graph is the maximum eccentricity or equivalently the longest shortest path. That is,

$$diameter = \max_{v \in V} \max_{u \in V} d(u, v).$$

- The radius of a graph is the minimum eccentricity. That is,

$$radius = \min_{v \in V} \max_{u \in V} d(u, v).$$

- The girth of a graph is the length of the shortest cycle. A cycle is a path that connects a node with itself.

9.4.2 Visualization

As stated above, traditional graph depictions such as node-link diagrams are widely used every day. However, the legibility of these depictions is limited as the graph grows larger and more complex.

9.4.2.1 Quilts and Geneaquilts

Quilts [2] are interactive matrix based depictions for layered graphs designed to address these problems. As the graphs grow larger, node-link diagrams have crossing edges issues while matrices take a lot of space but quilts provide a compact version of relationships. However, quilts primary challenge are their depiction of skip links, links that do not simply connect to a succeeding layer. It addresses this

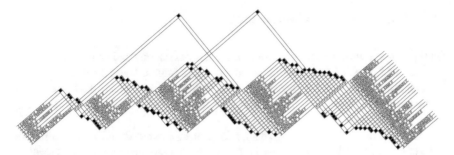

Fig. 9.7 A part of the royal family genealogy. Adapted from Geneaquilts [4]. Each *F* icon represents a family consisting of parents and children

issue by reinforcing color and other display elements (e.g., text) and making it inter-active to manage the path-finding task. The researchers compared the effectiveness of quilts using the most effective skip link depiction to node-link diagrams and centered matrices. The studies vary the number of nodes, links and skip links, and record time and accuracy as participants find paths. Eventually, as the graph complexity grows, quilts enable participants to maintain better performance than node-link diagrams and centered matrices. Furthermore, quilts have been applied to a genealogical appli-cation which is called Geneaquilts [4], a specialization of quilts that takes advantage of the bipartite nature of genealogical graphs. See Fig. 9.7.

9.4.2.2 Integration of Methods

Many researchers have attempted to integrate multi-dimensional visualizations with scatterplots and parallel coordinates [22, 24], instead of multi-view methods to per-form visual analysis. In fact, scattering points performs well in parallel coordinates when the distribution is distinguishable and meaningful by taking the advantages of both scatterplots and parallel coordinates. It may be useful in finding more sensi-tive variables than others so that the researchers can focus on the more important variables in the model [22].

Bezerianos et al. [3] exploit scatterplot matrices and node-link diagrams to visu-alize a multi-dimensional graph. Especially, they focus on the graph attributes from the space of edges, and objects with its degree, centrality, and clustering coefficients. The users can interact with the display to navigate through multiple projections of the data set.

Similarly, Viau et al. [23] introduce a graph-based interface which enables select-ing features within a multi-dimensional data set and compare graph metrics. They integrate a scatterplot matrix, a node-link diagram, and transitions to parallel coor-dinates.

9.5 Summary and Conclusions

In this chapter we have seen examples of tools for extracting and analyzing information in three types of data: documents, images and graphs. There are a large number of techniques to deal and analyze these types of data and, as stated in the introduction, there are quite a few other types of complex data. Key-value databases and streaming data, among others.

Analysis of complex data is related to context-enhanced and soft information fusion (e.g., [20, Chap. 14]). That is, fusion of data streams that are not based on sensor data and can be delivered out of the current time scope (e.g., weeks later). In this case, the complexity stems not only from combining different types of data, but also from the problem of using the same context for analysis as well as how to resolve conflicting data. In these situations, ensemble-based techniques [17] for combination of data can be employed, where the techniques addressed in this chapter can serve to obtain fingerprints or digests that can be employed by basic classification mechanisms.

Combination of data based on complex data types is also more complex than non-complex data types, since we cannot employ standard techniques such as principal component analysis for dimensionality reduction directly on the complex data.

References

1. Aggarwal, C. C., & Zhai, C. X. (2012). *Mining text data*. Springer Science & Business Media.
2. Bae, J., & Watson, B. (2011). Developing and evaluating quilts for the depiction of large layered graphs. *IEEE Transactions on Visualization and Computer Graphics (TVCG / InfoVis11)*.
3. Bezerianos, A., Chevalier, F., Dragicevic, P., Elmqvist, N., & Fekete, J. D. (2010). Graphdice: A system for exploring multivariate social networks. In *Proceedings of Eurographics/IEEE-VGTC Symposium on Visualization (Eurovis 2010)*.
4. Bezerianos, A., Dragicevic, P., Fekete, J.-D., Bae, J., & Watson, B. (2010). Geneaquilts: A system for exploring large genealogies. *IEEE Transactions on Visualization and Computer Graphics (TVCG / InfoVis10)*.
5. Bifet, A., & Gavaldà, R. (2007). Learning from time-changing data with adaptive windowing. In *Proceedings of the SIAM International Conference on Data Mining*.
6. Blei, D. M. (2012). Probabilistic topic models. *Communications of the ACM, 55*(4), 77–84.
7. Blei, D. M., Ng, A. Y., & Jordan, M. I. (2003). Latent dirichlet allocation. *Journal of Machine Learning Research, 3*:993–1022.
8. Choi, Y., Lee, H., & Irani, Z. (2016). Big data-driven fuzzy cognitive map for prioritising it service procurement in the public sector. *Annals of Operations Research*.
9. Dahl, G. E., Sainath, T. N., & Hinton, G. E. (2013). Improving deep neural networks for LVCSR using rectified linear units and dropout. In *2013 IEEE International Conference on Acoustics, Speech and Signal Processing (ICASSP)* (pp. 8609–8613). IEEE.
10. Duarte, D., & Ståhl, N. (2018). Machine learning. In A. Said, & V. Torra (Eds.), *Data science in practice*. Springer.
11. Friendly, M., & Denis, D. (2005). The early origins and development of the scatterplot. *Journal of the History of the Behavioral Sciences, 41*(2), 103–130.
12. Griffiths, T. L., & Steyvers, M. (2004). Finding scientific topics. *Proceedings of the National Academy of Sciences, 101*(suppl 1), 5228–5235.

13. Grn, B., & Hornik, K. (2011). topicmodels: An R package for fitting topic models. *Journal of Statistical Software, Articles, 40*(13), 1–30.
14. Inselberg, A. (1985). The plane with parallel coordinates. *Visual Computer, 1*(4), 69–91.
15. Kim, G.-H., Trimi, S., & Chung, J.-H. (2014). Big-data applications in the government sector. *Communications of the ACM, 57*(3), 78–85.
16. Krizhevsky, A., Sutskever, I., & Hinton, G. E. (2012). Imagenet classification with deep convolutional neural networks. In: *Advances in neural information processing systems* (pp. 1097–1105).
17. Polikar, R. (2006). Ensemble based systems in decision making. *Circuits and Systems Magazine, IEEE, 6*(3), 21–45.
18. Salton, G., & Buckley, C. (1988). Term-weighting approaches in automatic text retrieval. *Information Processing & Management, 24*(5), 513–523.
19. Simonyan, K., & Zisserman, A. (2014). Very deep convolutional networks for large-scale image recognition. arXiv:1409.1556.
20. Snidaro, L., Garcia, J., Llinas, J., & Blasch, E. (Eds.). (2016). *Context-enhanced information fusion: Boosting real-world performance with domain knowledge.* Cham, Switzerland: Springer. OCLC: 951075950.
21. Sonka, M., Hlavac, V., & Boyle, R. (2014). *Image processing, analysis, and machine vision.* Cengage Learning.
22. Steed, C., Shipman, G., Thornton, P., Ricciuto, D., Erickson, D., & Branstetter, M. (2012). Practical application of parallel coordinates for climate model analysis. In: *International conference on computational science, data mining in earth science.*
23. Viau, C., Mcguffin, M. J., Chiricota, Y., & Jurisica, I. (2010). The FlowVizMenu and parallel scatterplot matrix: Hybrid multidimensional visualizations for network exploration. *IEEE Transactions on Visualization and Computer Graphics.*
24. Yuan, P., Guo, H., Xiao, H., Zhou, H., & Qu, X. (2010). Scattering points in parallel coordinates. *IEEE Transactions on Visualization and Computer Graphics, 15*(6), 1001–1008.
25. Zeiler, M. D., & Fergus, R. (2014). Visualizing and understanding convolutional networks. In: *European conference on computer vision* (pp. 818–833). Springer.

Chapter 10
Big Data Programming with Apache Spark

Elio Ventocilla

In this chapter we give an introduction to Apache Spark, a Big Data programming framework. We describe the framework's core aspects as well as some of the challenges that parallel and distributed computing entail. No statistical background is required and neither are any other data analysis skills. It is, however, encouraged for the reader to be familiarized with a functional programming language (e.g. Scala) or with the concept of lambda functions (anonymous functions)—for these are used across most examples. Spark is built on the Scala programming language,[1] hence Scala is the language of choice for the examples given.

10.1 Introduction and Overview

Spark is a framework for simplified, distributed and parallel data processing and querying. In other words, it abstracts the complexity of distributed and parallel data computations, so that you can more easily make faster operations on your data. It can be deployed on a personal computer as well as on clusters (two or more computers working together). Some reasons that make Spark increasingly popular are:

- It leverages from in-memory storage (RAM) for faster access to data. When it cannot (because data is bigger than memory) it leaks to hard drive without losing what has been computed so far.

---

[1] See "Scala: From a Functional Programming Perspective" [3] for a quick introduction to the programming language; and/or "Programming in Scala" [2] for a more thorough description of the language.

E. Ventocilla (✉)
University of Skövde, Skövde, Sweden
e-mail: elio.ventocilla@his.se

© Springer International Publishing AG, part of Springer Nature 2019   171
A. Said and V. Torra (eds.), *Data Science in Practice*, Studies in Big Data 46,
https://doi.org/10.1007/978-3-319-97556-6_10

- It can be deployed on a personal computer providing easy access for learning purposes. When deployed on a personal computer it makes use of CPU cores for parallel computing.
- It supports four programming languages: Scala (language in which Spark is written in), Java, Python and R. It could also be said to support SQL-like queries, as later shown.
- It supports parsing and loading different data formats such as plain text, CSV, JSON, Parquet and others—whether compressed (e.g. gzip) or not. It also supports connections to relational databases (e.g. MySQL) and to NoSQL databases (e.g. Cassandra).
- It enables the possibility to create standalone applications, do interactive querying through the command line, as well as processing data streams.
- It provides a Machine Learning library built to leverage from Spark's engine which makes it scalable to large amounts of data.

These are some of the benefits of using Spark. It is also good, however, to know its limitations. Spark might not be appropriate as:

- A transactional database. Spark is a batch processing framework which is not suitable for inserting, updating or deleting single rows of data. This is better handled by SQL or NoSQL databases.
- A cluster manager. Spark does provide a standalone deployment mode (i.e. it can be setup on a cluster without a cluster manager such as Hadoop YARN), however, it does not provide all the benefits of a dedicated cluster manager such as resource management. This can come in handy if you are deploying other applications within the same cluster.

10.1.1 Installation

As of the writing of this book, the latest version of Spark is 2.2.0. We will not describe how to install it since the project web page provides a thorough—and up to date—guideline on how to do so. The installation package can be found at http:// spark.apache.org/downloads.html. Installation documentation can be found at http:// spark.apache.org/documentation.html.

As previously said, no cluster is needed to install and try Spark. Current distributions can be deployed on personal computers with Linux-based operating systems, Windows and Mac.

10.1.2 Quick Hands On

The quickest way to see Spark in action (after installation) is to open a terminal (command prompt), change directory to the Spark home and type:

```
user@his:~$ spark-shell
Using Spark's default log4j profile: org/apache/spark/log4j-defaults.properties
Setting default log level to "WARN".
To adjust logging level use sc.setLogLevel(newLevel). For SparkR, use setLogLevel(newLevel).
17/06/14 14:16:06 WARN NativeCodeLoader: Unable to load native-hadoop library for your platform... using
17/06/14 14:16:06 WARN Utils: Your hostname, Event resolves to a loopback address: 127.0.1.1; using 212.2
17/06/14 14:16:06 WARN Utils: Set SPARK_LOCAL_IP if you need to bind to another address
17/06/14 14:16:14 WARN ObjectStore: Failed to get database global_temp, returning NoSuchObjectException
Spark context Web UI available at http://127.0.0.1:4040
Spark context available as 'sc' (master = local[*], app id = local-1497442567597).
Spark session available as 'spark'.
Welcome to
      ____              __
     / __/__  ___ _____/ /__
    _\ \/ _ \/ _ `/ __/  '_/
   /___/ .__/\_,_/_/ /_/\_\   version 2.1.1
      /_/

Using Scala version 2.11.8 (Java HotSpot(TM) 64-Bit Server VM, Java 1.8.0_74)
Type in expressions to have them evaluated.
Type :help for more information.

scala>
```

Fig. 10.1 Output on a terminal after running the spark-shell script. Bold words (Web UI, sc, local[*] and spark) have been purposely highlighted

./bin/spark-shell

This will output some messages and leave you at scala> (see Fig. 10.1) where you can start typing code as in the Scala REPL.[2] If Python is your language of choice, then use ./bin/pyspark instead. In this context, using a REPL for making ad hoc queries on your data is referred to as interactive querying. The alternative to it are self-contained applications (an example is given in Sect. 10.5.1).

Highlighted in bold font in Fig. 10.1 we find:

- Web UI: a monitoring interface that can be accessed using the given URL (in our example is http://127.0.0.1:4040). There it is possible to see Spark's environment configuration as well as ongoing, completed and failed computations. We will not go through the UI interface in this short introduction to Spark.
- Spark context (sc): this variable is the main entry point to Spark's functionalities. Through it, Resilient Distributed Datasets (see Sect. 10.3), accumulators and broadcast variables can be created. The sc variable is automatically created with the spark-shell (and pyspark) command. We will work with it in the examples following in this chapter.
- Master (local [*]): the master refers to the environment where Spark is running, which, in this case, is the local computer. The asterisk specifies the amount of CPU cores that will be used when working in a local setting. This number can be set using the master option[3] e.g.

```
spark-shell --master local[4]\\
```

When working on a cluster the master should be set to a cluster URL e.g.

```
spark-shell --master mesos://HOST:PORT
```

---

[2] Read, Eval, Print and Loop. See http://docs.scala-lang.org/overviews/repl/overview.html.
[3] See more options by running spark-shell --help.

- Spark session (`spark`): this variable is an entry point to the Dataset and DataFrame API, which is an optimized way of working with tabular data. We give some examples after working with RDDs.

Our first example loads Spark's README.md file (hosted in Spark's home directory) using the Spark context variable:

```
val rdd = sc.textFile("path/to/README.md")
```

An instance of a Resilient Distributed Dataset (RDD) is created with the name `rdd`, which references the README.md file that, in this case, exists in the local file system. If you are working on a cluster managed by e.g. YARN, then accessing a file would be the same but with a cluster URL (e.g. `hdfs://path/to/file`). An RDD is a collection whose elements are distributed in a cluster. The elements of our `rdd` are strings, where each string is a line in the README.md file. As with other collections in Scala such as `Lists`, RDDs have operations for working with its elements. For example:

```
rdd.first // res0: String = # Apache Spark
rdd.count // res1: Long = 104
```

The `first` method returns the first element in the RDD which, in this case, is the first line in the README.md file. The `count` method returns the number of elements in the RDD which in our example is the number of lines in the README.md file. It is also possible to read directories as follows:

```
val rdd = sc.wholeTextFiles("path/to/directory")
```

Here the RDD will not be a collection of strings but a collection of tuples, where each tuple has one string element with the name of a file and one string element with the whole content of the file. Calling the `count` method on such RDD would return the number of files in the directory. Before diving into more elaborated examples of what can be done, let us have a look at the overall picture of Spark's architecture.

10.2   Architecture

Figure 10.2 represents an outline of how the framework looks like in a cluster. Spark is software which optimizes computations on a cluster architecture. There are hardware concepts of a cluster that are good to know so as to tell them apart from Spark's own software concepts. A *worker* or node is a computer dedicated to data processing and/or file storage. A worker, like any computer, has a *CPU* with one or more cores (C1, C2, etc.), a *RAM* memory, and a *hard drive* (HD)—or solid-state drive (SSD). A set of workers placed in a tower-like structure is called a *rack*. Figure 10.2 depicts three racks with five workers each, and each worker with their own CPU, RAM and HD. This type of hardware architecture tells us something about levels of speed access to data:

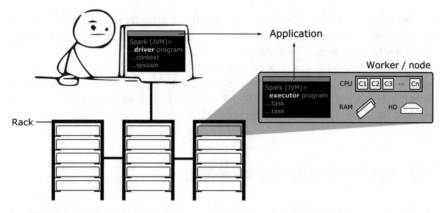

Fig. 10.2   The Spark software architecture when working on a cluster

1. The first and fastest (leaving cache memory aside) level of access to data is RAM. Any data taken from the RAM within the same computer will be retrieved at a much faster rate than at any of the levels described below.
2. The second fastest level of access is that from the HD. In spite of being the second fastest level, HD can be more than 10.000 times slower than RAM [1]. An SSD can improve performance compared to HD but would still remain slower than RAM.
3. The third level of speed access is within workers in the same rack, for example, a given node A reads data from another node B, where both A and B are in the same rack. In other words, access that requires network traffic.
4. Finally, the fourth speed level is access to data within workers from different racks. This can be seen as the final level assuming that the cluster is working solely in a local network.

Why know this? It is all about performance. When running a query or a Machine Learning algorithm on petabytes of data you would probably like to wait seconds rather than minutes or hours. Having a picture on where performance bottlenecks can happen gives a better hint on where improvements can be made. This also illustrates why Spark is fast: it leverages from RAM access as much as possible.

Having seen the hardware architecture of a cluster, we now move to the Spark's software architecture concepts: the executor programs, the driver program, and the application. The **executor** programs—one for each worker/node—are the ones in charge of carrying out jobs (computations). In this sense, jobs in Spark are physically divided among workers and logically managed by their executors (a job is further divided by executors into logical pieces called tasks which are units of work for handling data). The **driver** program coordinates the computations carried out by the executors. The Spark context and Spark session variables live in the driver. Finally, a Spark **application** is a term used to refer to a single driver program and its executors. It is important to note that, in a cluster setting, the driver and the executors all run on different workers. This is exemplified better later on (in Sect. 10.3.1) but it basically means that references to variables created in the driver are not shared among executors. In other words, be careful with variables!

When working on a personal computer the driver and an executor will run on the same machine sharing resources. It might also be the case they both run within the same Java Virtual Machine (JVM), thus making it tricky at times to prove the previous statement on non-shared variable references.

10.3   Resilient Distributed Datasets

The main abstraction through which Spark achieves reliable computations across workers is called Resilient Distributed Dataset (RDD).[4] Distributed because data is partitioned[5] into more than one physical—or virtual—workers/nodes; and resilient because they can recover computations from node failures as later shown in this section. RDDs can be seen as a simple Scala collection with the main difference that it abstracts the complexity of performing parallel computations of distributed data. Its class definition is RDD[T], where T denotes the type of elements the collection contains (e.g. String, Double, Array, etc). The following is equivalent to our first code example:

```
import org.apache.spark.rdd.RDD
val rdd: RDD[String] = sc.textFile("path/to/README.md")
```

RDDs are immutable[6] structures like Scala List, and they provide two types of operations over data: transformations and actions.

Transformations are lazy operations over the data i.e. operations that are not executed until needed or, in this case, until an action is called. These types of operations result in a new RDD being created, where the new RDD has in its "plan" all transformations from the old RDD plus the transformation that created it. Two examples of transformations[7] are map and filter. Let us do a word count example over the README.md file:

```
val rddWords = rdd.flatMap(_.split("␣"))
val rddLongWords = rddWords.filter(_.size > 3)
```

---

[4]A more technical description of RDDs can be found in [4].

[5]The number of partitions in which a file is divided is, if not stated otherwise, decided by Spark based on file block size. File block size is 32MB on a local file system, and 128MB on YARN. The minimum number of partitions is 2, which would be the case of small files such as README.md (3.8 K).

[6]Immutability is a key concept in functional programming, and an important aspect for reliable parallel programming.

[7]List of transformations: http://spark.apache.org/docs/latest/rdd-programming-guide.html# transformations.

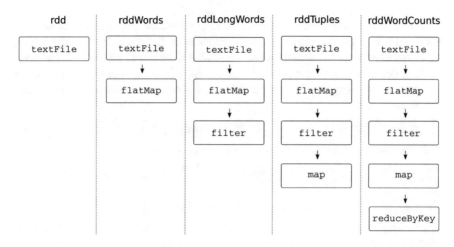

Fig. 10.3 Lineage (or DAG) for each RDD transformation

```
val rddTuples = rddLongWords.map((_, 1))
val rddWordCounts = rddTuples.reduceByKey(_ + _)
```

Step by step, our code: flat maps file lines into words by splitting them by spaces; filters (keeps) those words with more than 3 characters; maps the remaining words into tuples in the form of (<word>, 1); and reduces the collection by key (first element of the tuple) by adding ones (second element of the tuple).

None of these transformations have actually taken place. All we have done is to create a plan of execution (Fig. 10.3)—or **lineage**[8] in the form of a directed acyclic graph (DAG). The lineage is the key for RDDs' resilience: if a node fails in the middle of a computation, another node takes over using the lineage as a guideline to recompute whatever was lost.

Since each transformation returns a new RDD, the code can be rewritten as a chain of calls so as to have a more compact view. The following is equivalent to the previous:

```
val rddWordCounts = rdd
  .flatMap(_.split(" ")) // RDD[String]
  .filter(_.size > 3) // RDD[String]
  .map((_, 1)) // RDD[(String, Int)]
  .reduceByKey(_ + _) // RDD[(String, Int)]
```

Actions are operations that translate to actual computations. Two examples of actions[9] are count and first. When an action is called on an RDD, the lineage/DAG is sent to the executors in order to start computing all transformations in parallel. When transformations are done, partial results from each executor are—

---

[8]Lineage is the official name given in the Spark documentation.

[9]List of actions: http://spark.apache.org/docs/latest/rdd-programming-guide.html#actions.

depending on the action—sent back to the driver where the final output is computed. Some actions might not require partial computations to be sent back to the driver; some, for example, can request for processed data to be stored within each executor.

As a continuation of the previous code snippet, we can compute the number different words—or space-separated chunks of characters—with more than three characters length:

```
rddWordCounts.count // = 240 words
```

Or can we also take the first five most frequent words:

```
rddWordCounts
  .sortBy(_._2, false)
  .take(5) // (Spark,16), (using,5), (build,4),...
```

The `sortBy` call is a transformation that orders the list of words based on their counts (`_._2` i.e. second element of the tuple) in a descending manner (`false` parameter). A full list of operations, both transformations and actions, can be found in the RDD API documentation.[10]

10.3.1 Implications of Distributed Computations

Doing distributed computations with Spark implies that, among other things: some transformations might be recomputed unnecessarily if not stated otherwise (i.e. if not **persisted**); all operations will run on different executors; variables and methods might be encapsulated (i.e. **closure**) and sent to the executors in order to run the lineage; data might have to travel from one worker to another in order to make groupings and aggregations (i.e. **shuffle**).

The first implication regards **persisting** results in memory in order to avoid recomputations. Our last two code snippets computed `rddWordCounts`'s whole lineage, twice: once when we called `count` and a second time when we called `take(5)`. In order to avoid this and have these transformations computed only once, we have to tell Spark to persist a lineage transformations into memory. This is done by calling `persist`:

```
rddWordCounts.persist()
```

This will lift a flag telling Spark to save in memory the results of `reduceByKey` (which is the final transformation in `rddWordCounts`'s lineage) when it is computed for the first time (when calling `count`). The persistence flag can be seen as the only mutable attribute of RDDs. To lower the flag call `unpersist()`.

For the second implication, having Spark hiding the complexity of the underlying architecture, we might forget that some commands will not behave the same as if working on a personal computer. For example, the `println` command.

---

[10]http://spark.apache.org/docs/latest/api/scala/index.html#org.apache.spark.rdd.RDD.

```
rddWordCounts
  .foreach(line => println(s"Words:_${line.size}")
```

Running this code on a simple Scala collection such as a `List` would output all 240 tuples in the terminal. However, our collection is an RDD, meaning that the `println` command is run on different executors and workers. This code would not give any output on the driver and not, therefore, on the terminal where the command was given. This code, nonetheless, outputs to the `stdout` of each executor—which could be the intended instruction.

Another example is reading files or directories. It is not enough for the driver to have access to the path given through e.g. `sc.textFile("path")`; the workers also need access to the path given by the driver in order to load the file and run transformations over it—e.g. access to a shared network drive or an Amazon S3 bucket.

The third implication regards a process called **closure**. When an action is called on an RDD, variables and methods declared on the driver, and which are used within the RDD operations, are wrapped up in a JAR file and sent to the executors. This wrap up process is called closure, and what it means is that variables declared on the driver will not be the same as the ones sent to the executors—their initial values will be the same, but updates will happen on different variables. Say that, for example, instead of calling `rddWordCounts.count` to get the number of different words, we decide to use an auxiliary variable called `count`:

```
var count = 0
rddWordCounts.foreach(_ => count += 1)
count // = 0
```

The `count` variable, which was declared in the driver, will remain unchanged after running this code. `count` variables sent to the executors will be updated but we will not be able to see them. If shared variables (across executors) are needed, then have a look at **accumulator** and **broadcast** variables.[11] The following snippet yields the intended result:

```
val accum = sc.longAccumulator
rddWordCounts.foreach(_ => accum.add(1))
accum.value // = 240
```

Finally, the fourth implication is about a process called **shuffle**.[12] In our word count example, when performing the `reduceByKey` operation, we might find that instances of a word (e.g. Spark) exists in more than one worker (see Fig. 10.4). In order to perform this operation, Spark needs to compute word counts within each worker and then, taking these partial results, compute the overall word count from all workers by moving data from one to another. The process of moving data is called shuffle and transformations such as `reduceByKey` are called shuffle operations.

---

[11]http://spark.apache.org/docs/latest/rdd-programming-guide.html#shared-variables.

[12]http://spark.apache.org/docs/latest/rdd-programming-guide.html#shuffle-operations.

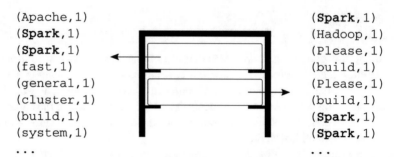

```
(Apache,1)                                    (Spark,1)
(Spark,1)                                     (Hadoop,1)
(Spark,1)                                     (Please,1)
(fast,1)                                      (build,1)
(general,1)                                   (Please,1)
(cluster,1)                                   (build,1)
(build,1)                                     (Spark,1)
(system,1)                                    (Spark,1)
  . . .                                         . . .
```

Fig. 10.4 A rack of two workers with a dataset partitioned among both. Together they conform the README.md file after the last *map* transformation. The Spark word is found on both machines

These types of operations are expensive and time consuming because they require not only network traffic, but also disk I/O and data serialization.

Using shuffle operations is probably inevitable but we should be conscious about it and use them when fewer data needs to be moved around. The following code is a suboptimal version of the previous because the shuffle happens before the filtering—which shuffles the whole dataset instead of the filtered chunks:

```scala
val rddWordCounts = rdd
  .flatMap(_.split(" "))
  .map((_, 1))
  .reduceByKey(_ + _)
  .filter(_._1.size > 3)
```

10.3.2 Types of RDDs

This is a brief overview of a couple of special RDDs: Pair and Double RDDs.

Pair RDDs are those whose elements are tuples in the form $(<K>, <V>)$, where K denotes a key and V a value. Our word count example worked with a pair RDD, where the key was a word and the value a 1. This type of RDDs provide special functions for traversing and grouping elements by their key element e.g. reduceByKey, groupByKey, foldByKey, join and others.[13]

Double RDDs are those whose elements are of type Double. This type of RDDs give access to functions such as mean, variance, histogram and others.[14] A way to compute the mean frequency of words would be:

---

[13]http://spark.apache.org/docs/latest/api/scala/index.html#org.apache.spark.rdd. PairRDDFunctions.

[14]http://spark.apache.org/docs/latest/api/scala/index.html#org.apache.spark.rdd. DoubleRDDFunctions.

```
rddWordCounts
  .map(_._2) // RDD[Double]
  .mean
```

10.4 Datasets and DataFrames

Datasets are abstractions over RDDs. They provide better performance (in terms of processing times) with (hopefully) less coding effort for the programmer/analyst. This is achieved thanks to a so called Catalyst Optimizer which, in a broad sense, takes high level queries and assembles RDD transformations in an optimized way.

Datasets are collections which contain either elements of specific Scala data types (e.g. String, Double, Array) or elements of a case class. DataFrames, on the other hand, are an alias to Datasets with elements of type Row; that is, collections of elements of type Any (i.e. Row[Any]). Objects from Datasets can, in contrast to Rows in Dataframes, be serialized[15] in a way that does not require their later deserialization for certain operations (e.g. sorting and filtering), thus further improving performance.

DataFrames are Datasets with elements of type Row. Row objects are basically collections of any type of elements (Row[Any]). DataFrames provide less information about the type of data that is to be transformed, therefore optimizations from the Catalyst Optimizer become limited. Datasets, on the other hand, provide the types of data they hold (e.g. String, Double, a case class, etc.) so the Catalyst Optimizer can perform further optimizations.

Loading data as a DataFrame can be done in two ways: using the spark variable or from existing RDDs. Let us do an example with a simple CSV dataset that looks like this:

```
date,amount
2016-06-30,-169.79
2016-06-28,-55.70
2016-06-27,-56.30
...
```

We load this data into a DataFrame using the spark variable as follows:

```
val dfExpenses = spark
  .read // DataFrameReader
  .option("header", true)
  .option("inferSchema", true)
  .csv("/path/to/expenses.csv")
```

The DataFrameReader[16] allows us to read different file formats such as JSON, Parquet, etc. For CSV files we can give some options/guidelines to the reader so that

---

[15]Serialization occurs when data is sent over the network e.g. when a shuffle operation takes place.
[16]http://spark.apache.org/docs/latest/api/scala/index.html#org.apache.spark.sql.
DataFrameReader.

it can parse the file better: the "header" option states whether the first line of the file contains column names; the "inferSchema" option requests the parser to try and determine columns' data types (e.g. double, string, date, etc.). If the delimiter between columns in your data is other than commas, then use `option("delimiter", " ")`. We can see the schema by calling:

```
dfExpenses.printSchema
```

```
root
 |-- date: timestamp (nullable = true)
 |-- amount: double (nullable = true)
```

We can see that Spark has inferred the type of data we have. If we had left the options out, the schema would look like this:

```
root
 |-- _c0: string (nullable = true)
 |-- _c1: string (nullable = true)
```

Calling the `show` method prints out the first ten records of the DataFrame in a structured manner:

```
dfExpenses.show
```

```
+--------------------+--------+
|  date|  amount|
+--------------------+--------+
|2016-06-30 00:00:...|  -169.79|
|2016-06-28 00:00:...|  -55.7|
|2016-06-27 00:00:...|  -56.3|
...
```

Querying DataFrames has a similar syntax to that of SQL. For example, we can select those expenses higher than 100—i.e. transactions below −100—and sort them from highest to lowest:

```
dfExpenses
  .select($"amount")
  .where($"amount" < -100)
  .orderBy($"amount")
```

The dollar sign in front of strings is syntactic sugar for instantiating a `Column`[17] object for the given column name. Another way to get columns is by using the DataFrame variable instead:

```
dfExpenses.select(dfExpenses("amount"))
```

---

[17]http://spark.apache.org/docs/latest/api/scala/index.html#org.apache.spark.sql.Column.

Column objects gives us access to expression methods such as less than (<), equal (===), plus (+), and many others. The following is an example of an average aggregation:

```
dfExpenses
  .groupBy(month($"date").as("month"))
  .agg(avg($"amount"))
  .orderBy($"month")
  .show
```

```
+-----+------------------+
|month|  avg(amount)|
+-----+------------------+
|    1|-319.1927450980392|
|    2|-117.9860465116279|
|    3| -232.450243902439|
. . .
```

The functions month and avg are provided by Spark as a part of an off-the-shelf set of operations[18] for Datasets.

Let us do a more elaborated example with a weather data set[19] from the NOAA institution. We will parse the data using RDDs and transform them into Datasets in order to perform aggregations. In that sense, we need a case class to represent the structure of each record in the data set:

```
case class Temperature(
  ts: java.sql.Timestamp, // recorded timestamp.
  value: Double, // temperature.
  quality: String // temperature measurement quality.
)
```

There are many more fields that can be taken from the dataset but, for simplicity purposes, we use only three. Next step is to create functions to parse string records to Temperature objects:

```
val dateFormat = new java.text.SimpleDateFormat("yyyyMMddhhmm")

def stringToTimestamp(s: String): java.sql.Timestamp =
  new java.sql.Timestamp(dateFormat.parse(s).getTime())

def stringToTemperature(s: String): Temperature =
  Temperature(
    stringToTimestamp(s.substring(15, 27)),
    s.substring(87, 92).toDouble,
    s.substring(92, 93)
  )
```

---

[18]http://spark.apache.org/docs/latest/api/scala/index.html#org.apache.spark.sql.functions$.

[19]Weather from 2012: http://academictorrents.com/details/16be344abd95d58afd4860445f4a927b7eb1a89d.

By using these functions it is now fairly straightforward to load data into a
`Dataset`:

```
val ds = sc
  .textFile("/path/to/weather/NOAA")
  .map(stringToTemperature) // RDD[Weather]
  .toDS
```

Now we can make the following query: retrieve the average, minimum and maximum recorded temperatures per month, where measurement quality is 1 (good):

```
ds
  .where("quality_=_1") // equivalent to $"quality" === "1"
  .groupBy(month($"ts").as("month"))
  .agg(avg($"value"), max($"value"), min($"value"))
  .orderBy("month")
  .show
```

```
+-----+-------------------+----------+----------+
|month|  avg(value) |max(value) |min(value) |
+-----+-------------------+----------+----------+
|   1|  43.67431049241436|  450.0|  -597.0|
|   2|  62.22730350319569|  467.0|  -553.0|
|   3|  87.10457422783658|  450.0|  -464.0|
. . .
```

Many of the grouping and aggregating functions provided by Spark
accept columns names as strings e.g. `avg("value")` as an alternative to
`avg($"value")`. Not all, however, accept this, so it might be convenient to just
stick to `$` column instantiation.

10.5 Streaming

Streaming regards handling data that arrives continuously. A common example is that
of tweets from Twitter. The current version of Spark is able to deal with streaming
data in micro batches i.e. small sets of data. There are two ways of working with
streams: using RDDs and using DataFrames (called structured streaming). We will
start with RDDs.

10.5.1 RDD Streaming

When dealing with streams a new actor takes part in the Spark software architecture: receivers. These are a special type of executors that are only concerned with
assembling batches of data into RDDs for other executors to process (see Fig. 10.5).

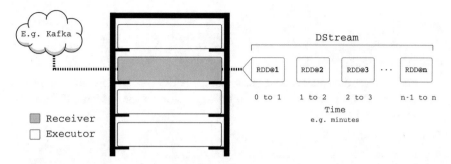

Fig. 10.5 Spark streaming outline. The `Receiver` listens to incoming data and produces RDDs with partitions of data distributed across conventional executors for their processing

The collection of RDDs produced by the receiver (or receivers) is called Discretized Stream (`DStream`).[20] `DStreams` are our data type abstraction through which streams of data are handled.

There are two types of transformations that can be done over `DStreams`: stateless and stateful. The former are transformations that only take into account data from the last assembled batch (i.e. newly arrived data); the latter, on the other hand, are transformations that also take into account data from previously assembled batches (i.e. both new and old data). We begin with an example of stateless transformations.

Let us say we have access to live weather data from the devices that provide data to the National Oceanic and Atmospheric Administration (NOAA) agency. The recorded data arrives every 10 s to a repository to which we have access. As in our `DataFrame` example, we are interested in the average temperature but, in this case, for each 10 s window. For our streaming examples let us use self-contained applications.[21] The basic project file structure is the following:

```
<project-folder>
| build.sbt
| src
  | main
    | scala
      | App.scala
```

The `build.sbt` file defines project library dependencies.[22] For the streaming example, we need `spark-sql` and `spark-streaming`. The sbt file should look like:

```
name := "RDD streaming"
version := "1.0"
```

---

[20] A `DStream` is collection of `RDDs`, which are collections of distributed elements. This might sound confusing but bear with us.

[21] How to build self-contained applications: http://spark.apache.org/docs/latest/quick-start.html#self-contained-applications.

[22] Library dependencies can be found in the Maven repository: https://mvnrepository.com/.

```
scalaVersion := "2.11.8"

libraryDependencies += "org.apache.spark" %% "spark-sql" % "2.2.0"
libraryDependencies += "org.apache.spark" %% "spark-streaming" % "2.2.0"
```

Through these dependencies we get access to Spark's core and streaming libraries. The rest of our work will take place in the `App.scala` file, which will have the following structure:

```
import org.apache.spark.sql.SparkSession
import org.apache.spark.streaming._

object App {
  def main(args: Array[String]) {
    // Rest of our code goes here.
  }
}
```

In Scala, unlike in Java, the name of the class/object does not have to be the same as the name of the file. It is, however, a good practice for them to be the same. In this case we named them both `App`, but we might as well have named them `HelloWorld` (just keep the name in mind when deploying). Now we can start doing actual Spark streaming programming.

Within our application we create a `spark` session variable and, with its Spark context, we instantiate a streaming context (`ssc`) variable:

```
val spark = SparkSession
  .builder
  .appName("Streaming_example")
  .getOrCreate()

val ssc = new StreamingContext(spark.sparkContext, Seconds(10))
```

Unlike in the REPL, in self-contained applications we need to instantiate the `spark` session variable ourselves. The streaming context variable (`ssc`) is like the Spark context variable in the sense that it provides an entry point for creating collections (`DStreams` in this case). When creating it we need to specify the frequency with which batches of data will be computed (e.g. every 10 s). Before continuing with the main streaming code, we will reuse/copy-paste classes and parsers used in the `Datasets` example:

```
case class Temperature(...)
val dateFormat = new java.text.SimpleDateFormat...
def stringToTimestamp(s: String): Timestamp = ...
def stringToTemperature(s: String): Temperature = ...
```

Through the `ssc` variable we instantiate an input `DStream`:

```
val streamInput = ssc.textFileStream("/path/to/weather/data")
```

The `streamInput` variable is of type `DStream[String]`, in spite of having RDDs underneath. These sort of collections can be handled almost like RDDs but they have some different transformations and actions. In the next step we map each record to an object of type `Temperature`, and filter them so as to keep only those with measurement quality of 1:

```
val streamTemperature = streamInput
  .map(stringToTemperature)
  .filter(_.quality == "1")
```

The only difference (code-wise) so far with RDDs and static data is the call `ssc.textFileStream`. Under the hood, however, things will be less alike: with an RDD all data is mapped and filtered at once; with DStreams, mapping and filtering is applied to incoming data every batch interval (e.g. 10 s, as defined in the instantiation of the `ssc` variable) for as long as we let it.

Moving on, we will now compute some statistics over our cleaned, structured data. For that purpose we make use of the following helper `case class`:

```
case class Statistic
  (sum: Double, count: Int, min: Double, max: Double) {

  def +(other: Statistic): Statistic =
    Statistic(
      sum + other.sum,
      count + other.count,
      min.min(other.min),
      max.max(other.max))

  def avg(): Double = sum / count

  override def toString(): String =
    s"count:_$count,_min:_$min,_max:_$max,_avg:_$avg"
}
```

The `case class` is used as a structure for holding the sum of temperatures, their count, and their minimum and maximum values. It has a method + for "adding" another `Statistic` object, which will become convenient in the `reduce` transformation later. The `avg` method computes the division between `sum` and `count`; and the `toString` method is overridden in order to have an output to console that includes the average. Let us now define the operations for computing the statistics for every batch:

```
val streamStats = streamTemperature
  .map(w => Statistic(w.temp, 1, w.temp, w.temp))
  .reduce(_ + _) // DStream[Statistic] with count = 1
```

Using `reduce` on a DStream returns a new DStream with a single element as a result of the reduce operator (`_` + `_`). In that sense, the reduce operation is

a transformation and not an action as in RDDs. In order for Spark to compute the
defined transformations we need, therefore, an action. In this case, we use `print`
to display results in the console/terminal:

```
streamStats.print()
```

The `print` method belongs to `DStream` type objects and it is not the same as
the inbuilt one from Scala. The final required steps for stream computations to run
are the following:

```
ssc.sparkContext.setLogLevel("WARN") // optional
ssc.start()
ssc.awaitTermination()
```

Setting the log level to "WARN" keeps Spark's logging output to a minimum,
making it easier to read the results. To run our application we first have to package
it using `sbt` in the command line, inside the project folder:

```
sbt package
```

And then submit it to Spark as follows:

```
spark-submit \
  --class "App" \
  --master local[*] \
  target/scala-2.11/rdd-streaming_2.11-1.0.jar
```

This outputs the following example result every 10 s, where each output represents
statistics for the last 10 s window of newly arrived data:

```
-------------------------------------------------
Time: 1501684810000 ms
-------------------------------------------------
count: 90323, min: -430.0, max: 442.0, avg: 36.354...
```

In this first example we used: files as a streaming source; *stateless* transformations
that do not take previous batches into account; and `print` in order to display results
in console. There are other streaming sources to which Spark can connect[23] such as
sockets; and also other ways to handle results such as e.g. saving them as text files:

```
streamStats.saveAsTextFiles("path/and/prefix", "suffix")
```

In order to compute transformations that take into account results from previous
batches (i.e. **stateful** transformations) we have two alternatives: windowed transformations and the `updateStateByKey` methods.

All windowed transformations take (at least) the following two parameters:
`windowLength` and `slideInterval`. The former basically defines the number of RDDs to be computed every given slide interval. Both, `windowLength` and
`slideInterval`, are to be given in time units that are divisible by the batch time

---

[23]http://spark.apache.org/docs/latest/streaming-programming-guide.html#basic-sources.

frequency defined in the streaming context. In our example, the batch frequency is 10 s, hence window and slide can be e.g. 10, 20, 30 s, etc. If batch frequency is 5 min, then window and slide can be 5, 10, 15, ... minutes. Following our streaming example, we can compute the following window operation:

```
streamTemperature
  .map(w => Statistic(w.temp, 1, w.temp, w.temp))
  .reduceByWindow(_ + _, Seconds(30), Seconds(20))
  .print()
```

The `reduceByWindow` operation can be translated to "sum elements (of type `Statistic`) in windows of 30 s, every 20 s", or "sum elements of every 3 batches (RDDs), every 20 s". Sum every 3 batches (or RDDs) because we are requesting 30 s windows over 10 s batch frequency (defined in the instantiation of the streaming context `ssc`). Having 20 s slide interval will mean that the last RDD in our first window will also be the first of our second window (plus two newly arrived), and so on.

Windowed operations will take into account previous batches but they will not truly keep aggregations over time (e.g. statistics per week over the course of a year). For this –when using RDD streaming– we have to implement a function that will handle the aggregates for every batch. Let us do an example for computing temperature statistics per week of the year. Mainly we define a function that will handle the aggregates, which we named `updateState`:

```
def updateState(
  newValues: Seq[Double],
  state: Option[Statistic]): Option[Statistic] = {

  val oldState = state match {
    case Some(s) => s
    case None => Statistic(0, 0, Double.MaxValue, Double.MinValue)
  }
  val newState = newValues
    .map(t => Statistic(t, 1, t, t))
    .foldLeft(oldState)(_ + _)

  Some(newState)
}
```

This function takes two parameters: the first one (`newValues`) is a sequence of newly arrived elements (in this case temperatures) corresponding to a given key (in this case week of the year); the second (`state`) represents the aggregate computed from previous batches/sequences (in this case temperature statistics of the past 3 batches) for the same key. The actual key is not of our interest in this case so we just trust that Spark will provide the correct parameters per key. In the body of the function we first retrieve the old state (`case Some(s)`). If no previous state exists (`case None`), we instantiate a new one. Thereafter we map `Temperature` into `Statistic` objects and do a `foldLeft` using the `oldState` as an initial value. This provides an aggregate of the newly arrived temperature values, with the old aggregate/state from previous batches. Finally, we return this value as `Some` (which, like `None`, is a subclass of `Option`), so that Spark can provide it in the next call as the parameter `state`; and so on.

For aggregating statistics per week of the year we need a pair `DStream` (similar to pair `RDDs`), with elements in the form of `(Int, Double)`. The first element of type `Int` will represent the week of the year, and the second of type `Double`, the temperature. For this purpose we define the following function to map `Temperature` objects to tuples:

```
def temperatureToTuple(t: Temperature): (Int, Double) = {
  val c = java.util.Calendar.getInstance()
  c.setTime(t.ts)

  (c.get(java.util.Calendar.WEEK_OF_YEAR), t.value)
}
```

With these two functions, `updateState` and `temperatureToTuple`, we can now define our stateful transformation:

```
weatherStream
  .map(weatherToTuple) // DStream[(Int, Double)]
  .updateStateByKey(updateState _)
  .print()
```

This type of stateful transformations can only be done on pair `DStreams`. Moreover, in order to guarantee resilience, these transformations require that Spark has a place to make checkpoints (backups). This can be any path executors have access to:

```
ssc.checkpoint("/path/to/checkpoint")
```

Running our code would yield the following example output:

```
---------------------------------------------
Time: 1501762640000 ms
---------------------------------------------
(52,count: 913, min: -249.0, max: 300.0, avg: 63.216...)
(4,count: 669, min: -319.0, max: 293.0, avg: -32.18...)
(16,count: 671, min: -92.0, max: 390.0, avg: 83.3815...)
...
```

10.5.2 Structured Streaming

Structured streaming was added in Spark 2.1 and, as of the writing of this book, has undergoing improvements. The intention is to provide developers with an API that works consistently with `Datasets`, regardless of the source (static or streaming). Let us redo the weather streaming example from scratch using `Datasets`. In the self-contained application we start by importing the following (`build.sbt` remains the same):

```
import org.apache.spark.sql.SparkSession
import org.apache.spark.sql.functions._
import org.apache.spark.sql.streaming.ProcessingTime
import scala.concurrent.duration._
```

As before, SparkSession will provide the means to create the spark variable; functions._ will give us access to aggregation functions (e.g. avg, min, etc.); and the last two imports will be used to define how often computations should be triggered (i.e. batch windows). In the RDD streaming example we did not give importance to where case classes were defined as long as they were accessible to our self-contained application. In this case, however, we have to define them outside of the object where the main function is defined:

```
case class Temperature
   (ts: java.sql.Timestamp, temp: Double, quality: String)

object App {
  def main(args: Array[String]) {
    ...
  }
}
```

This is because of a bug in Spark that fires when it tries to encode the contents of a DataFrame object, so as to have a typed Dataset. Do not worry much about it. Just leave case classes that are used to define the type elements of DataFrames outside of the object scope. What follows does go within the main function.

```
val spark = SparkSession
  .builder
  .appName("Structured_streaming")
  .getOrCreate()

import spark.implicits._
```

When doing structured streaming we do not need a streaming context variable as in the RDD streaming example. The spark session variable will do. The implicits import is needed so that Spark can encode contents of DataFrames to e.g. String to Temperature objects. We again reuse the following parsing code:

```
val dateFormat = new java.text.SimpleDateFormat...
def stringToTimestamp(s: String): Timestamp = ...
def stringToTemperature(s: String): Temperature = ...
```

And now we can write the core code for structured streaming:

```
val streamTemperature = spark
  .readStream
  .textFile("/path/to/weather/data")
  .map(stringToTemperature)
  .filter(_.quality == "1")
```

This code creates a `Dataset` with `Temperature` elements that have a measurement quality of 1. The only difference with a static data source comes from the `.readStream` command. If we use `read` instead, we tell Spark that data is static. Other types of file streams can be loaded. To do so, instead of calling `textFile`, use `parquet`, `json` or `csv`. If you wish to connect to a socket then you need to do something like the following:

```
val streamSocket = spark
  .readStream
  .format("socket")
  .option("host", "localhost")
  .option("port", 9999)
  .load()
```

More examples on input sources can be found in Spark's structured streaming guide.[24]

Since the `streamTemperature` object is a `Dataset` it can be treated as such:

```
val dsStatistics = streamTemperature
  .groupBy(weekofyear($"ts").as("week"))
  .agg(
    count($"temp").as("count"),
    min($"temp").as("min"),
    max($"temp").as("max"),
    avg($"temp").as("avg"))
```

Here we compute the same statistics as in the RDD streaming example. The `weekofyear` function is provided by Spark. We now have to define an action or "sink" to which computed data should be poured into.

```
val queryStream = dsStatistics
  .writeStream
  .trigger(ProcessingTime(10.seconds))
  .outputMode("complete")
  .format("console")
  .start

queryStream.awaitTermination()
```

Now we have some new elements:

- The `trigger` call defines the batch processing frequency. This is equivalent to the frequency defined in the instantiation of the `ssc` variable, for RDD streaming.
- The `outputMode` call defines what is to be "poured" into the sink. In this case, `"complete"` means that the whole table contained in `dsStatistics` will be displayed in console. Two other options are `"append"` and `"update"`. The

---

[24]http://spark.apache.org/docs/latest/structured-streaming-programming-guide.html#input-sources.

former means that results from new batches are added to results from previous batches. This would be the common output mode when doing non-aggregative transformations—i.e. for simpler operations such as map and filter—because no old values are being updated (e.g. update statistics for week 3, when we are already in week 4, but some data arrived late). It is also the common mode when pouring the results to the file system because, at each trigger, new files are created, which Spark cannot update. The latter output mode, "update", states that only updates on old records (e.g. the update week 3 statistics) should be outputted to the sink.

- Finally, the format call defines the "sink" i.e. where data is to be poured into. Options are "console", "memory" and file formats e.g. "parquet", "json", etc. Console, as the name states, prints out to console. Memory, on the other hand, creates a temporary table in the driver's memory—so use with caution—which can be queried through the Spark session variable (e.g. spark.sql("select * from <table name>")). For this to work it is compulsory to provide a query name (or table name) to which queries can be made e.g. queryName("temperature"). For file format sinks it is necessary to provide folder and checkpoint locations through the option method e.g. option("path", "/path/to/folder") and option("checkpointLocation", "/path/to/other/folder"). One last "sink" is to call forEach, in which case, through a ForeachWriter object, you define what to do with the transformed data.

Leaving our code as it is, we are ready to deploy it. We compile it using sbt and then submit it using spark-submit, as we had done in the RDD streaming example. The output, in this case, will have a more structured format:

```
------------------------------------------------
Batch: 0
------------------------------------------------
+----+-----+------+-----+------------------+
|week|count|  min|  max|  avg|
+----+-----+------+-----+------------------+


| 31|64388|-695.0|517.0|  216.5921755606635|
| 34|65513|-665.0|490.0|205.55906461313023|
...
```

The complete code for streaming examples seen here can be found in Github as weather-streaming.

10.6   Other Big Data Tools

There are many other programming frameworks that help dealing with large datasets and/or streaming data. As with Spark, these frameworks seek to abstract the complexities of parallel and distributed computing, while reliably boosting data crunching tasks.

We will give a brief description to each, leaving the details to other dedicated documentations.

- **Apache Flink** is focused on streaming, real-time data. A difference with Spark streaming is that it handles data elements as soon as they arrive i.e. it does not assemble batches. This is, however, an aspect of Spark that seems to be improving as of the writing of this book.[25] In general Flink comes with two promises: lower latency (i.e. faster) and better memory management. If latency is a critical issue, then probably Flink is the solution for you.
- **Apache Kafka** is also focused on streaming, real-time data. It provides tools for creating data processing pipelines as well as interfaces through which external apps can publish or subscribe to data streams.
- **Apache Lucene** deals with text data. It provides indexing for fast, high-level querying over large text corpora. Built on top of Lucene you find other popular solutions such as **Elasticsearch** and **Apache Solr**, each with their own pros and cons.
- **Logstash** aims at making it easy to congregate and transform data in different formats and from multiple sources. Similar, but maybe less popular, is **Solr**.

This is far from being an exhaustive list of frameworks and it is only meant to provide an idea of what is out there. To choose the right solution depends greatly on the needs of the problem at hand.

References

1. Jacobs, A. (2009). The pathologies of big data. *Communications of the ACM, 52*(8), 36–44.
2. Odersky, M., Spoon, L., & Venners, B. (2011). *Programming in Scala*, 2 edn. Artima Press.
3. Torra, V. (2016). *Scala: From a functional programming perspective: An introduction to the programming language.* Cham, Switzerland: Springer.
4. Zaharia, M., Chowdhury, M., Das, T., Dave, A., Ma, J., McCauley, M., Franklin, M.J., Shenker, S., & Stoica, I. (2012). Resilient distributed datasets: A fault-tolerant abstraction for in-memory cluster computing, in *Proceedings of the 9th USENIX conference on networked systems design and implementation, USENIX Association, Berkeley, CA, USA, NSDI'12*, p. 2

---

[25] A new feature called Continuous Processing, which will allow handling data elements as soon as they arrive, is being released.

Author Index

© Springer International Publishing AG, part of Springer Nature 2019   195
A. Said and V. Torra (eds.), *Data Science in Practice*, Studies in Big Data 46,
https://doi.org/10.1007/978-3-319-97556-6

Printed in the United States
By Bookmasters